Information Hiding in Speech Signals for Secure Communication

Information Hiding in Speech Signals for Secure Communication

Wu Zhijun

Science Press
Beijing

SYNGRESS

AMSTERDAM • BOSTON • HEIDELBERG • LONDON
NEW YORK • OXFORD • PARIS • SAN DIEGO
SAN FRANCISCO • SINGAPORE • SYDNEY • TOKYO

ELSEVIER

Syngress is an Imprint of Elsevier

Syngress is an imprint of Elsevier

The Boulevard, Langford Lane, Kidlington, Oxford OX5 1GB, UK
225 Wyman Street, Waltham, MA 02451, USA

First edition 2015

British Library Cataloguing in Publication Data
A catalogue record for this book is available from the British Library

Library of Congress Cataloging-in-Publication Data
A catalog record for this book is availabe from the Library of Congress

ISBN–13: 978-0-12-801328-1

For information on all Syngress publications
visit our website at http://store.elsevier.com/

14 15 16 17 18   10 9 8 7 6 5 4 3 2 1

Contents

Preface

In the information communication field, speech communication via network becomes an important way to transfer information. With the development of information technology, speech communication is widely used for military, diplomatic, and economic purposes as well as in cultural life and scientific research. Therefore, speech secure communication and the security of communication information have attracted more and more attention. The rapid growth of the Internet as both an individual and business communication channel has created a growing demand for security and privacy in the network communication channel. Security and privacy are essential for individual communication to continue and for e-commerce to thrive in cyberspace.

Any type of multimedia data—for example speech, transmitted via network—needs to be protected from manipulation, forgery, and theft. More sophisticated attacks require that more advanced security technologies be avoided, which have to be optimized for the particular requirements of each application scenario. As a development tendency of information security and a fresh-born technique, information hiding breaks the mentality of traditional cryptology. Information hiding technology carefully examines information security from a new perspective. Traditional secure communication approaches cannot satisfy current security requirements, and the garbled bits are likely to attract attention, or even encounter attacks from others. It is urgent to introduce a new mechanism for secure communication; thus the author undertakes it as his task to investigate a more secure communication, based on information hiding technology, which has many distinguishing characteristics. The most important characteristic is that there are univocal and continuous plain text speech signals on the communication line, so the secure communication will be more covert and safe.

OBJECTIVES

This book intends to develop the theory and system of real-time secure speech communication over a network based on information hiding technology, and to provide an overview of the research area of information hiding in secure communication. In the book, the author attempts to break the massive subject into comprehensible parts and to build, piece by piece, a development of information hiding technology in secure communication. The book emphasizes the fundamentals of algorithms and approaches concerning the technology and architecture of a secure communication field, providing a detailed discussion of leading-edge topics such as filter similarity and linear predictive coding (LPC) parameters substitution.

ORGANIZATION OF THE BOOK

This book introduces several methods to hide secret speech information in different types of digital speech coding standards. In the past 10 years, the continued advancement and exponential increase of network processing ability have enhanced the efficiency and scope of speech communication over the network. Therefore, the author summarizes his years of research achievements in the speech information hiding realm to form this book, including a mathematical model for information hiding, the scheme of speech secure communication, an ABS-based information hiding algorithm, and an implemented speech secure communication system, which are organized into different sections in accordance with the security situation of the network. This book includes nine chapters, which introduce speech information hiding algorithms and techniques (embedding and extracting) capable of withstanding the evolved forms of attacks.

The four parts of this book are as follows:

I. Introduction, includes Chapter 1. This book begins with an introductory chapter, where the approaches of secure communication and their realization are discussed in general terms. This chapter ends with an overview of speech coding standards. This part provides a source of motivation for interested readers to wade through the rich history of the subject. The concept and algorithms involved in this part are explained in details in subsequent chapters.

II. Theory, consists of Chapters 2 and 3. This part proposes a mathematical model, and embedding and extracting algorithms of secure communication based on information hiding. The design of a speech information hiding model and algorithm for secure communication focuses on security, hiding capacity, and speech quality. The model and algorithm may help readers to develop a deep understanding of what information-hiding-based secure communication is all about.

III. Approaches, consists of Chapters 4 through 7. This part presents a detailed embedding secret speech coded in MELP 2.4 K, with the secret speech coded in accordance with five important families of the speech coding standard, such as G.721, G.728, G.729, and GSM, which are used in Voice over Internet Protocol (VoIP) or Public Switched Telephone Network (PSTN). Next is the introduction of the process of extracting secret speech coded in MELP 2.4 K in the G.721, G.728, G.729, and GSM standards. The procedure of the realization of secure communication based on specific speech coding standards is developed.

IV. Realization, consists of Chapters 8 and 9. This part gives two applications of secure communication based on the technology of information hiding. Each chapter sets up one example to show how to implement secure communication over VoIP and PSTN by using the information-hiding-based model and algorithm individually.

Each chapter begins with a high-level summary for those who wish to understand the concepts without wading through technical explanations. Then examples and more details are provided for those who want to write their own programs. The combination of practicality and theory allows engineers and system designers to implement true secure communication procedures, and to consider probable future developments in their designs, therefore fulfilling the requirements of social development and technological progress for new types of secure communication.

CHARACTERISTICS OF THE BOOK

The author introduces readers to earlier research regarding security of transmitted speech information over networks, and combines information technology and speech signal processing to form a speech hiding model for solving the problem of secure speech communication. Taking into consideration the requirements of real time and security, the author puts forward the algorithm of LPC coefficients substitution for secret speech information hiding based on filter similarity (LPC-IH-FS) and information extraction with blind-detection-based minimum mean square error (BD-IE-MES). Compared with traditional secure communication, those proposed theories and algorithms have distinct advantages in the field of secure communication. Moreover, the chapters and sections of this book are arranged from the simple to the complex, and the chapters relate to each other closely, with a reasonable knowledge structure. The book is scientifically and systematically organized. Plenty of tables and diagrams work jointly to interpret how the approach works and to demonstrate superior performance of the proposed approach. In addition, to validate the performance of the proposed approach, plenty of experiments are conducted, and detailed experiment results and analyses are available in the book.

A few topics may be of great interest for readers, such as the new idea of speech secure communication, the introduced mathematical model for information hiding, the proposed LPC-IH-FS and BD-IE-MES algorithms, and the corresponding experimental results. The highlight of the book is that the author applied information hiding technology in secure communication for the first time and then addressed a new embedding and extracting approach based on the analysis-by-synthesis (ABS) algorithm.

NEW IDEA OF SPEECH SECURE COMMUNICATION

This new idea of speech secure communication is based on information hiding technology, which differs from the traditional speech secure communication. Conventional secure communication systems transmit encrypted or transformed noisy signals over the public channel. The proposed speech secure communication system in this book delivers clear, comprehensive and meaningful stego speech, with secret speech embedded, over the public channel. The stego speech is explicit but secret speech is implicit.

NEWLY PROPOSED SPEECH INFORMATION HIDING TECHNOLOGY

The ABS algorithm introduces speech synthesizing into speech coding, and speech embedding and coding are synchronously completed. Combined with conventional embedding approaches, ABS is capable of adjusting hiding capacity on the basis of secret speech. It may achieve a maximum hiding capacity of 3.2 kbps, under the condition of G.728 16 kbps carrier speech with MELP 2.4 kbps secret speech. The proposed approach has the advantages of high hiding capacity, good imperceptibility, and high embedding rate.

In the book, the embedding and corresponding extracting methods are illustrated with elaborate diagrams and tables. Analyses and comparisons are also given. The proposed method puts emphasis on solving the secure problems occurring in real-time speech communication, and it achieves a relatively high hiding capacity, thus the requirements of real-time communication are met.

The major contributions of this book lie in extending a new hotspot in a secure communication scope and opening up a perspective application realm of information hiding technology. The particular insight of information security and the unique ABS approach infuses this newborn technology with vitality.

POTENTIAL READERS OF THE BOOK

This book can be a valuable reference book for anyone whose profession is information security. Readers may be scientists and researchers, lecturers and tutors, academic and corporate professionals, even postgraduate and undergraduate students.

Readers will learn related concepts and theory about speech secure communication. This book offers readers in-depth knowledge of the theory, modules, algorithms, and systems about information hiding or secure communication proposed by the author. The knowledge allows potential customers to protect their secure speech communication against even the most evolved wiretapping and information analysis attacks. Furthermore, by combining the theory with practice, readers can not only conduct related experiments in accordance with the contents of the book, but also implement a secure communication system by programming. In short, the achievements presented in this book can be used for network security in practice.

To better understand this book, readers should have prerequisite knowledge in communication principles, networking theory, basic theory of information hiding, and speech signal processing technology.

Acknowledgments

Many people have contributed to the publication of this book.

I sincerely express my thanks to Professors Yang Yixian and Niu Xinxin, my mentors in the research area of information hiding at Beijing University of Posts & Telecommunications, for writing a section of the book, and for putting forward many valuable suggestions to improve the book.

I really appreciate my research team member Mr. Yang Wei at Beijing University of Posts & Telecommunications, who completed the analysis-by-synthesis (ABS) research and implementation of speech coding, and made a great contribution to the core algorithm in this book.

I would like to express my thanks to Associate Professor Ms. Ma Lan at Civil Aviation University of China, who made a great contribution to the information hiding model and analysis of speech coding standards.

I am truly indebted to Professor Wang Jian, Professor Zhang Yanling, Professor Yin Hengguang and Dr. Ma Yuzhao at Civil Aviation University of China, for improving the language of the book.

The input from anonymous reviewers of the book is also appreciated.

I want to very much thank my team members, lecturer Ms. Lei Jin and Mr. Yue Meng, for their strong support in every aspect.

I am grateful to my graduate students, Mr. Wang Chen and Ms. Liu Wanhui, for the simulation experiments (using MATLAB) in VoIP that are included in this book, and for their work on some sections of the book.

I thank my graduate students Ms. Cao Haijuan and Ms. Zhao Ting, for preparing the figures included in Chapters 2, 3, 8, and 9.

I thank my lab colleagues, Mr. Yang Wei, Dr. Bai Jian, and Dr. Yang Yu, for their kind permission to reproduce their figures in the book.

I am grateful to Ms. Chen Jing, Technical Editor of the Science Press on Information Technology, for helpful comments on the organization of the book, and for her patience and meticulous work on improving the book.

I thank the editor Mr. Zhang Pu at Science Press for getting in touch with me early, and helping me to complete the proposal and evaluation of the book.

I am indebted to my graduate students, Ms. Cao Haijuan and Ms. Ma Shaopu, for their help in checking the typos in all chapters and the references.

I thank Professor Ms. Han Ping, the Dean of School of Electronics & Information Engineering, Civil Aviation University of China, for approving and providing financial aid for the publication of the book.

I thank my colleagues in the School of Electronics & Information Engineering, Civil Aviation University of China, for their selfless assistance and sincere concern.

I also want to extend my sincerest thanks to everyone who supported me during my time as an author.

I am deeply grateful to the China National Science Foundation and Tianjin Natural Science Foundation. This work is partly financially supported by the China

National Science Foundation (No. 61170328), Tianjin Natural Science Foundation (No. 12JCZDJC20900, U1333116), the Fundamental Research Funds for the Central Universities-Civil Aviation University of China (CAUC) under grant 31122013P007, 3122013D003, and 3122013D007, the Civil Aviation Science and Technology Innovation Fund in 2013, the research laboratory construction funds of Civil Aviation University of China (CAUC) in 2014-2016, and the postgraduate courses construction funds of Civil Aviation University of China (CAUC) in 2013 under grant 10501034.

Overview

This book focuses on secure speech communication via VoIP (Voice over Internet Protocol) and PSTN (Public Switched Telephone Network) by using the technology of information hiding, an emerging steganographic subject. Secure speech communication is the practice of hiding secret speech digital information in public speech, requiring hiding capacity, robustness, and imperceptibility by means of steganographic techniques. This book addresses the key problems currently faced by communication security, and the technology under study facilitates communication information confidentiality and ensures the integrity and security of communication information. This book proposes the approach of analysis-by-synthesis speech information hiding (ABS-SIH) to hide secret speech into public/carrier speech for the purpose of secure communication. A number of different types of speech coding schemes used in VoIP and PSTN, such as G.711, G.721, G.728, G.729, and GSM, are applied to realize secure communication by using the proposed algorithms of linear predictive coding (LPC) parameter substitution for secret speech information hiding based on filter similarity (LPC-IH-FS) and secret speech information extraction with blind detection based on minimum mean square error (BD-IE-MES). Each scheme is described in a number of different operational environments, such as VoIP and PSTN.

Also included in the book are details of the new emerging and synthetic technology-information hiding, which aims at hiding useful or important messages in other information to disguise the existence of the messages themselves. The research primarily focuses on the fundamentals of information theory, applying information hiding theory and technology in real-time speech communication. This text not only puts forward the mathematical model used for speech information hiding, but also designs and realizes five concrete schemes based on different speech coding standards. It is more important that the book proposes LPC-IH-FS speech embedding and BD-IE-MES extracting algorithms, and develops a real-time speech secure communication system based on speech information hiding technology with DSP (digital signal processor) arrays.

The contents of this book include:

- Proposing a constraint speech secure communication model based on the theory and technology of information hiding.
- Presenting a secret speech information embedding algorithm LPC-IH-FS by using LPC parameters substitution based on filter similarity.
- Putting forward a secret speech information extracting algorithm BD-IE-MES with blind detection based on MES.
- Proposing an approach of information embedding and extracting for concrete speech coding schemes, such as G.711, G.721, G.728, G.729, and GSM, by using LPC-IH-FS and BD-IE-MES algorithms.
- Presenting an approach of secure communication over VoIP based on matrix coding.
- Bringing forward a scheme of real-time secure communication via PSTN based on the technology of information hiding.

The proposed LPC-IH-FS and BD-IE-MES algorithms guarantee information is transmitted through VoIP and PSTN secretly. They use the characteristics of LPC in the ABS coding method, choose different speech coding schemes (for example, G.711, G.721, G.728, G.729, and GSM) as the public speech carrier and the Mixed-Excitation Linear Predictive (MELP) 2.4 K scheme as secret speech, thus realizing the security of speech information. When comparing the secure communication method based on information hiding technology with the traditional secure communication method, results show that the proposed method is more efficient than the traditional secure communication method in terms of security and speech quality.

The real-time secure speech communication using the technology of information hiding is an innovative approach. This approach combines information security technology and emerging communication technology, opening up a new research field for studying the method of secure communication. Meanwhile, this approach makes it possible to explore a new application for information hiding technology in the communication field.

Introduction

Information communication is one of the most significant features of an information society. As an important part of information communication, secure communication protects state secrets, commercial secrets, and personal privacy, which is crucial for the nation, society, and individuals. Human beings are living in an information society, and communication security and confidentiality are used not only for military purposes, but also for public life, such as network voice communication, mobile communication, electronic payment, and mobile banking on the Internet [1].

There are many kinds of methods for secure communication over networks, although the available methods have different levels of shortcomings [2]. New approaches of secure speech communication are proposed to transmit security information via Voice over Internet Protocol (VoIP) and Public Switched Telephone Network (PSTN) based on the techniques of information hiding discussed in this chapter.

1.1 BACKGROUND

With the development of network communication technology, speech communication technology has seen a gradual transition to VoIP communication from the original PSTN network communication [3]. At present, the speech communication network has become a hotspot in international and domestic telecommunications development. More and more speech services through networks are realized.

As we know, network-based threats have become more sophisticated, and PSTN and VoIP calls are vulnerable to threats such as session hijacking and man-in-the-middle attacks [2]. Without proper protective measures, attackers could intercept a PSTN or VoIP call and modify the call parameters or addresses (numbers). Even without modifying PSTN numbers or VoIP packets, attackers may be able to eavesdrop on telephone conversations being carried over a PSTN or VoIP network. If VoIP packets are traveling unprotected over the Internet, the attackers have the opportunity to access the information that these packets carry [4–7].

With a standard PSTN or VoIP connection, intercepting conversations requires physical access to telephone lines or access to the private branch exchange (PBX) and switch or router. Speech or data networks, which typically use the public Internet

Information Hiding in Speech Signals for Secure Communication. DOI: 10.1016/B978-0-12-801328-1.00001-X and the TCP/IP protocol stack, need safeguards to protect data security and personal privacy. In this context, the research of speech secure communication is particularly important [8].

1.1.1 PROGRESS IN SECURE COMMUNICATION

The traditional speech secure communication technology mainly adopts the methods of Analog Scrambling [9,10] and Digitized Encryption [11,12]. Analog Scrambling is performed by segmenting and then scrambling the speech signals in the frequency domain, time domain, or both domains at the same time in order to change the intelligible speech signals into unintelligible signals. In Digitized Encryption, the speech signal is digitalized and then encrypted. Both of these traditional approaches rely on the transformation of speech signals themselves and cryptographic strength to protect the security communication of secret information. In recent decades, cryptographic strength has continuously improved. By contrast, the development of speech secure communication systems has been slow. The encryption system of Analog Scrambling has poor security performance, but with the development of voice signal digital technology, Digitized Encryption is widely used in speech secure communication [13,14].

At present, there are many successful encryption algorithms in modern cryptography technology, and their confidentiality is extremely high. So far, the most successful method of secure communication is implemented by data encryption, which has a certain degree of security and is easy to implement. But this method has inherent shortcomings [15,16].

First, most of the current encryption algorithms are designed through specific calculation procedures. The strength of encryption algorithms depends on the computing capacity of the computer in use, and the improvement in the rate of undecipherable passwords relies on increasing the length of the key. With the rapid growth of the computing power of computers, password security is always faced with new challenges.

Second, the approach transforms plain text into cipher text by using an encryption algorithm, and then the cipher text is transferred to the open channel. The encrypted data stream is almost random and meaningless gibberish, implying the encrypted information is important and secret, and attracts attackers' attention. By monitoring the communication channel, attackers can easily identify the encrypted information. Once gibberish is intercepted, the target of the attack will be found. Various kinds of attack methods can be used for cryptanalysis.

Third, in the case of a small amount of secret information exposure, attackers can gain valuable information by analyzing the characteristics of disclosure information and tracing the content of secret information, even though attackers cannot decode the messages to gain the entire secret content.

The traditional secure telephony based on an encryption system has the same drawbacks [17]. The transferred speech information is hidden in the cipher text. For the eavesdropper, the intercepted information has a high degree of ambiguity, or it can be seen as a kind of interference or noise. Undoubtedly, the communication performance of such a secure communication system can be designed perfectly, but its security performance might not be optimal. Once the secure communication lines were monitored and detected by eavesdroppers, two actions the eavesdroppers may take are:

1. Intercepting secret information from the communication channel carefully and deciphering the information meticulously.
2. Destroying the secure communication line and preventing the transfer of confidential information.

Moreover, the encryption method of the traditional secure telephone is in the time domain or frequency domain, which has two shortcomings:

1. The encryption algorithms and methods are limited, and most of them are based on the analog signal. Encrypted information is easily detected and destroyed.
2. The process of encryption and decryption may lead to different degrees of decline in the quality of the original speech signal.

In addition, a relatively new type of secure communication is chaotic secure communication [18,19]. This approach mainly makes use of the chaos system's sensitivity to the initial state and parameter dependence to protect the security of confidential information communication. Although chaotic secure communication has been a breakthrough in theoretical and experimental studies, there are still more practical application issues left, and many practical problems such as system synchronization need to be solved [20].

Whenever multimedia data—for example, speech—are transmitted via a network, they need to be protected from manipulation, forgery, and theft. More elaborate attacks demand more sophisticated security technologies, which have to be optimized for the particular requirements of each application scenario. A newly developed technique of information security, information hiding, breaks the traditional cryptology thought concepts. Information hiding technology carefully examines information security from a new perspective [21]. Traditional secure communication approaches cannot satisfy the current security requirements, because the garbled bits are likely to attract attention or even attacks from others. It is imperative to introduce a new mechanism for secure communication, which is why we address this issue. Compared with traditional secure communication and chaotic secure communication, secure communication based on information hiding technology has several distinctive characteristics for secure communication. The most important characteristics are univocal and continuous plain text speech signals that are transmitted on the communication line, so secure communication will be more covert and safer [22].

Traditional secure communication depends mainly on the strength of encryption algorithms. However, with the advent of the quantum computer, computing power has been increased, and many encryption algorithms are easily broken. Therefore, traditional secure communication is faced with serious challenges, leaving them in a difficult situation.

The emergence of the chaos approach brings secure communication to life. Chaotic secure communication has been in use for only 10 years [18]. As an important part of chaos application, chaotic secure communication has made remarkable progress not only in principle but also in practical application. Particularly, many new ideas and methods have been proposed in recent years. The new algorithm has achieved remarkable progress in practical and security aspects, and promotes chaotic secure communication as a practical technology. Chaotic secure communication is a new emerging interdisciplinary subject. It is still in the development stage, although there have been many research achievements. Most of the work has been performed in an ideal environment by replacing traditional communication carriers with chaotic sequences. Cryptographic pseudonumber computer simulation is another area of study. At present some chaos-theory-based patents are at an experimental prototype stage, and cannot implement a real chaotic secure communication system yet [23].

The author compiles his publications using one of two algorithms: the LPC parameters substitution for secret speech information hiding based on filter similarity (LPC-IH-FS) [24], and information extraction with blind detection-based minimum mean square error (MES-BD) [25]. The proposed algorithm adopts the speech synthesizer in the speech coder. Speech embedding and coding are synchronous; that is, the fusion of secret speech information data and speech coding. It can embed dynamic secret speech information data bits into original carrier speech data, with good efficiency in steganography and good quality in output speech. It is practical and effective compared with the existing methods, thus real-time communication requirements can be met.

1.1.2 A NEW TECHNIQUE FOR SECURE COMMUNICATION: INFORMATION HIDING

Although the idea of information hiding (or steganography) appeared very early, this technology was not gradually systematized until the 1990s [21,22]. After that, information hiding penetrated into various application fields, especially since the 1st International Information Hiding Science Conference 1996, London [21,22]. Then this kind of information technology received a lot of attention in research and industrial manufacturing, and made great progress.

Information hiding is a newly emerging information security technology, and it has been brought into play in many fields. More and more digital video, audio signals, and images are "pasted" with invisible labels, which carry hidden copyright marks to prevent unauthorized duplication [26]. Military systems widely use information security technology to encrypt secret messages. Moreover, the transmitter, receiver, and even the message itself are masked by using information hiding technology. Similar technologies are also used in mobile phone systems and other electronic media systems.

1.1.2.1 Basic Concepts of Information Hiding
The main idea of information hiding derives from ancient steganography, more strictly called "information cover," which literally means "covered writing" or "handwriting

cover." Information hiding can also be called "information concealing" or "information veiling." In summary, steganography aims at hiding useful or important messages into other information to disguise their existence [21,22].

As the name implies, information hiding technology embeds secret messages into host (or public) information with a certain transparency, which seems quite normal. Here, the secret messages and public documents (that can also be information) can be any kind of digital media, such as images, speeches, videos, common texts, and so on.

Hostile attackers cannot judge whether the transmitted information contains secret information. The host information that contains secret messages does not attract attention or cause suspicion. Similar to biological camouflage where animals can avoid being found by their enemies, the possible risk of danger is minimized. This is the essence of information hiding technology.

Information hiding technology integrates several research subjects and technologies in different fields, such as communication, cryptography, network signal processing, and voice and image coding. It is an advanced technology with strong comprehensive multidisciplinary theory [21,22]. It embeds a message (called a secret or confidential message) into another message (called public information or carrier) by utilizing the human sense redundancies for digital signals. As a result of the hiding procedure, a synthetic message presents its external feature, thus the basic characters and use value of the carrier (or public information) are preserved. The biggest advantage of information hiding is that except for the two communicating sides, any other third party cannot sense the existence of a secret message. Compared with traditional encryption algorithms, the extra protection layer makes the secret messages *invisible* rather than *incomprehensible* [21,22].

There is one thing that must be pointed out: traditional information security mainly focuses on the strength of the cipher algorithm; steganography mainly concentrates on concealing the secret message. These two views are not contradictory, but it is a good complementary relationship. The difference between them is that the different applications have different requirements. In practical application, they can be used as a complementary technique, existing at the same time. For example, to encrypt secret messages before embedding is a better way to ensure the security of information. This preprocessing method is more functional in practical applications, which means the coexistence of encryption and information hiding.

1.1.2.2 Basic Principles and Classifications

Information hiding technology comprises a large number of research areas. It may be divided into several aspects in accordance with Fabien A. P. Petitcolas's classification [21], as shown in Figure 1.1.

The classification of information hiding is explained next [21,22].

Covert Channel

Built on the common channel, the covert channel is a logical channel for sending covert messages. Actually, this channel does not exist in reality, it only utilizes information hiding technology to build a logical channel over a public channel. The

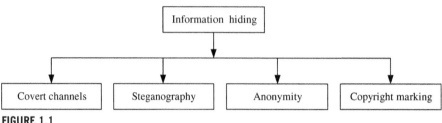

FIGURE 1.1

Classification of information hiding technology.

establishment of the covert channel is key for an information hiding system. Once the covert channel is found or destroyed, then the whole information hiding system will be paralyzed.

Steganography

Steganography is the general name of secure communication methods, and refers to the technology that embeds secret messages into public information. The so-called public information is a kind of information that does not cause any attention from others. The hiding method usually depends on the hypothesis that the third party cannot sense the existence of the covert communication. This method is used mainly in point-to-point covert communication between two parties that trust each other. Therefore, steganography is not robust. For example, the covered information could not be restored after data change.

Anonymity

Valid Internet users may employ the anonymous communication mechanism to ask for help or to vote secretly in an online election. But no one is willing to provide such a mechanism, because either intentional or unintentional attackers can easily overload it.

Copyright Marking

The main impetus for the development of information hiding is the copyright concerns. The audio, video, image, or other works can be in digitized form and it is very easy to obtain a perfect duplication. Publishers of music, movies, books, software, and other media are most concerned with the emergence of various unauthorized duplicates. Recently, the most significant study includes digital watermarks (hidden copyright information) and digital fingerprints (hidden serial number). The fingerprint is used to identify the copyright violators and the watermark is used to prosecute them.

These four categories are interrelated. The present research focuses on steganography and copyright marks. Specifically, technology-based image steganography and digital image watermarking are the hot points [21,22].

Digital Watermarking

Embedding special information in digital works (such as still images, video, audio, etc.) can certify copyright ascription and trace copyright infringement. The special information may be the author's serial number, company logo, meaningful text, and so on. In contrast with the camouflage, the hidden information in watermark is robust enough to resist attacks. Even if the watermarking algorithm is public and the hidden information is known by attackers, it is still difficult for them to devastate the hidden watermark (impossible in the ideal situation). In cryptography, the well-known Kerkhoffs principle [16] goes like this: the encryption system is still safe even under the condition that the attacker knows the theory and algorithm but does not know the corresponding key. Robustness requires that watermarking algorithms embed less information in public data than camouflage. Watermarking and steganography are complementary rather than competing.

Data Hiding and Data Embedding
Data hiding and data embedding usually are used in different contexts, and generally refer to camouflage or applications between camouflage and watermarking. In these applications, there is no need to protect the embedded data that are open to the public. For example, embedded data may be auxiliary information or service; they are obtained for free, and have nothing to do with copyright protection and access control functions.

Fingerprinting and Labeling
Fingerprinting and labeling refer to the particular purpose of a watermark. The watermark consists of information about a digital work that belongs to a creator or purchaser, which is embedded into the media. Each watermark is a unique code in a set of codes. The information in the watermark identifies a unique copy of a digital product.

Steganography
Steganography is the art and science of encoding hidden secret information in a way that no one, apart from the sender and intended recipient, suspects the existence of this information. It is a form of security through obscurity.

Steganography includes the concealment of secret information within media files, such as image and speech, and communication protocol, for example Session Initiation Protocol (SIP).

Copyright Protection for Digital Media
This technique of information hiding provides a method of digital media copy protection by embedding copyright information into digital media files. The method includes a process of hiding digital media data with a public key, using a hybrid cryptographic technique, a process of watermarking the media data, and a measurement compliance testing process in an effective way.

1.1.2.3 Applications of Information Hiding
Information hiding technology has increasingly drawn attention from research institutions and industry. The main driving force comes from people's concerns about copyright. Along with the digitization of audio, video, images, and other products, it is much easier to duplicate unauthorized digital products. Unauthorized product duplication has garnered enormous attention from musicians, movie makers, book

writers, and software publishers, hence research on two important branches of information hiding—watermarking and fingerprinting—appeared. Watermarking can be used as legal evidence in a copyright dispute to accuse pirates, whereas fingerprinting can be used to trace the pirates [21,22].

In general, the applications of information hiding technology can be summarized in the aspects of data confidentiality, copyright protection, nonrepudiation, antifake, and data integrity. Details are as follows [21,22].

Data Confidentiality

The purpose is to prevent the transmitted data in a network from being intercepted by unauthorized users and to avoid attacks by malicious attackers. This is an important element of network security. With economic globalization, the network security issue involves not only political and military elements, but also commercial, financial, and personal privacy issues. As a solution, information hiding technology can be used to protect the information that must be delivered through the network. The information is sensitive data in e-commerce, secret agreements and contracts between the negotiators, sensitive information in online banking transactions, digital signatures for important documents, personal privacies, and so on. These data will be protected and do not arouse any interest from attackers. In addition, some contents that are unwilling to be known by others can be hidden, so that only the experts who use the recognition software will acquire the secret messages.

Copyright Protection

Protecting the copyright of electronic image products is a motivation to promote the development of information hiding technology. With the rapid popularization of networking and digital technique, people are provided an increasing number of digital services that are propagated through networks, such as digital libraries, digital publishing, digital television, and digital news. These services are digital works, which are easy to modify and duplicate. And these characteristics will enormously harm the service providers' benefits and obstruct the development and popularization of advanced technologies.

Nonrepudiation

When utilizing e-commerce for online transactions, neither side of the trade could deny the actions he or she made or the actions of the other party. The two sides of the transaction use information hiding technology to embed their own feature marks into transmitted messages. Obviously, the feature marks can be viewed as secret messages and the encrypted feature marks can be regarded as watermarks. This watermark must be irremovable, thus the purpose of confirming the behavior can be achieved.

Antifake

In order to confirm the authenticity of confidential files, antifake marks can be appended through information hiding technology. Confidential files may be important documents for the secret work units or institutions, maps issued by a publishing house, various forms of material in business activities, and so on.

Data Integrity

For data integrity verification, the main purpose is to confirm that the data had not been falsified while being transmitted over the Internet or in the stored procedure. It is very easy to distinguish the modified media, which are protected by fragile watermark technology. The watermark is fragile because the watermark would be destroyed during the information data change.

1.1.2.4 Characteristics of Information Hiding

Generally speaking, information hiding or steganography technology has the following characteristics [21,22]:

- **Imperceptibility**. This is a fundamental requirement of steganography. The media data should not be degraded overtly after going through a series of hiding procedures, and the hidden information should not be seen or heard.
- **Security**. The hidden information must be secure. The location where the hidden information is embedded should also be safe. At least the destruction will not be caused due to the information format conversion.
- **Symmetry**. Typically, in order to correspond with the sender and the receiver of the communication system, the embedding and extraction processes are symmetrical. Coding, decoding, encryption, and decryption are the core technologies used in these processes.
- **Error correction**. Error correction coding is used to ensure data integrity and to rebuild the destroyed information after operations or conversion.

The common characters are invisibility, immeasurability, and robustness. Invisibility means that the quality of original information of the secret message should not be degraded after the embedding procedure. It also requires that the external features of public information (or carrier) are not altered sharply. For example, the carrier may be images; the modification must not be perceived by the human visual system after images go through the synthesizer [21,22].

Immeasurability mainly fights against hostile third parties. The third parties can hardly detect that certain information does contains secret messages, hence they cannot get the secret messages. At least it is impossible for them to know the secret messages during the valid time.

Robustness has the meaning of steadiness. It requires that the embedding algorithm be stable. Thus the embedded secret information cannot be removed easily. If the composite data changed, the embedded message should reflect the change and maintain data integrity. The secret information should be detected with required probability.

1.2 INTRODUCTION TO SPEECH CODING

From the viewpoint of communication, coding is in fact a signal process. It transforms a signal into a suitable form that can be transmitted in a channel well. Therefore, in digital communication, speech coding has a close relationship with the digitalization of speech signal. Coding is generally divided into source coding and channel coding [27–29]. Source coding improves the signal transmission and storage efficiency, which refers to the bit-rate of compressed digital speech signal [30]. (In communications and computer networks, bit-rate, sometimes written bit rate or as the variable R, is the number of bits that are conveyed or processed per unit of time.) Making the same channel capacity transmit multiple signals, or decreasing the signal storage volume, is also known as speech signal compression coding [30]. Channel coding, sometimes called reliability coding [31], is a treatment to improve transmission reliability. The following discussions involve only compression coding of the speech signal.

1.2.1 BASIC PRINCIPLES OF SPEECH CODING

In digital communication systems, speech signals are encoded into binary sequences, stored or transmitted through channels, then decoded and recovered to understandable speech. Encoding speech signals into binary sequences for transmission or storage has unique advantages. For example, this method can avoid noise interference during transmission and storage procedures. It is well known that the noise from analog channels more or less distort the signals, while digital communication systems possess enough relay stations to get rid of noise interference. Another advantage is that it is easier to process, encrypt, regenerate, and forward encode signals. In recent years, the development of Very Large Scale Integrated circuits (VLSI) enabled the digit speech coding technology to obtain widespread application [32].

The simplest method for digital coding is the direct analog–digital conversion. As long as the sample rate is high enough and the bit amount for quantifying each sample is large enough, the useful information can be reserved and the quality of the recovered speech can be retained. However, it needs a high bit-rate to digitize analog speech signal directly. For instance, if ordinary telephone systems employ a sample rate of 8 kHz and 12 bits to quantify a sample, the bit-rate required to transport speech may approach 96 kb/s. Such a high bit-rate is not suitable for transmission in the general channel. In order to improve transmission efficiency, coding compression is necessary [33].

There are two commonly used compression means. One is to reduce the quantization bits for each voice sample, while maintaining relatively good voice quality. Since the process object is speech waveform, this method is called Waveform Coding technology [34]. Another method, Voice-code technology [35], analyzes signals first to extract a set of characteristic parameters that have carried the main information of speech signals. They require fewer bits for coding, and speech can be regained using these parameters after they are decoded. The decrease of bit-rate mainly depends on the analyzed and extracted parameters, as well as the type of synthesizer.

1.2.1.1 Redundancy of Speech Signals

There are two fundamentals for compression coding. First, a big redundancy existed in speech signals as a result of a speech signal generation mechanism and the properties of its structure. Subsequently, compression coding tries to identify and eliminate redundancy so as to decrease bit-rate. Generally speaking, redundancy manifests mainly in the following aspects [30]:

- There is a strong correlation between signal samples; that is, the short-time spectrum is not flat.
- The dull resonance speech segment is quasi-periodic.
- The shape of the sound track and its change speed is limited.
- The probability distribution of transmitted codes is uniform.

The first three redundancies are determined by the generation mechanism of the speech signal and the last one has something to do with particular coding algorithms.

Using human hearing is another way to compress speech. The audio masking phenomenon is a very important characteristic, which means that a strong sound can veil a simultaneously existing weak sound. The quantized noise within the signals may be inhibited by use of this characteristic [36].

1.2.1.2 Two Types of Coding Methods

Speech coding can be generally divided into waveform coding and analysis-by-synthesis (ABS) methods. In the waveform coding method, each sample value of rebuilt speech signal $\hat{s}(n)$ should be close to the sample value of original signal $s(n)$ [37–39].

Let

$$e(n) = \hat{s}(n) - s(n) \tag{1.1}$$

where $e(n)$ stands for quantization error or reconstruction error.

The purpose of waveform coding is to minimize the energy of error sequence $e(n)$ under a given transmission bit-rate. Therefore, the signal-to-noise ratio (SNR) is always a useful performance indicator in waveform coding. The ABS method [39] is based on a speech signal generation model. It transforms speech signals into speech parameters, which are used for coding. Thus it is also known as parameter coding. In ABS systems, the decoded synthesis speech and original speech do not have a strict correspondence relationship between the samples. Due to the lack of objective tests, the quality of synthesis speech needs subjective judgment [40].

1.2.2 SPEECH CODING STANDARDS

For the sake of military secure communication requirements, research on speech coding technology began in 1939 [41]. Homer Dudley, the originator of speech coding technology, while at Bell Telephone Laboratory, proposed and implemented a channel vocoder, which could be used for speech signal transmission in telephones and telegram cable [42]. The ITU-T (CCITT) announced the 64 kbit/s pulse code modulation (PCM) speech coding algorithm recommendation G.711 in 1972 [43], which is widely used in digital communications, digital switches, and other fields. In 1980, the American government announced the 2.4 kbit/s linear prediction coding algorithm LPC-10 [44], which made it possible for digital telephone signals to be transported in the regular telephone bandwidth. In the early 1980s, ITU-T started to research an under-64 kbit/s non-PCM coding algorithm, and then passed the 32 kbit/s ADPCM coding recommendation G.721 in 1984 [45]. This standard can achieve the same sound quality with PCM. Moreover it has better performance in antibit error. In 1988, the United States announced a 4.8 kbit/s code excited linear prediction (CELP) coding algorithm [46]. At the same time, Europe launched a 16 kbit/s regular pulse excited linear prediction coding (RPE-LPC) algorithm [47]. The sound quality of these algorithms can reach a higher level that is much higher than that of the LPC vocoder.

Since the 1990s, along with Internet global development, people began to focus on VoIP technology. Thus IP packet speech communication technology got a breakthrough in theory and practical applications. ITU-T published its 16 kbit/s low-delay code excited linear prediction (LD-CELP) recommendation G.728 [48], and it has been widely used in practical applications due to its low delay, low bit-rate, and high performance. It is optional for VoIP technology. In 1996, ITU-T issued 5.3/6.4 kbit/s standard G.723.1 [49], and then 8 kbit/s conjugate structure algebraic code excited linear prediction (CS-ACELP) coding recommendation G.729 was put forward at the ITU-T SG15 conference in November 1995 [50]. Six months later, Annex A of G.729 was approved, and complexity reduced 8 kbit/s CS-ACELP coding algorithm became an official international standard [51]. These types of speech coding have become optional algorithms for VoIP technology. Some of the other speech coding standards are listed in Table 1.1 [52].

The quality of synthetic speech is said to be the most fundamental indicator for evaluating speech coding performance. There are many methods to evaluate the quality of speech by comparing the performance of different coding standards. Methods proposed over the years can be summed up in two categories: subjective evaluation method and objective evaluation method [52–54], as explained next.

1.2.2.1 Subjective Evaluation Method

This method is based on testers' listening sensibility. Compared with the original speech, the distortion degree of synthesis speech is graded according to the preset criterion. This method reflects how people feel the sound. Three measurements are commonly used: Mean Opinion Score (MOS), Diagnostic Rhyme Test (DRT), and Diagnostic Acceptability Measure (DAM). At present, the most popular subjective evaluation method is MOS, which uses five grades of rating criteria [54]. Dozens of testers listen to the synthesis speech in the same channel environment, and then give the score. The average scores are shown in Table 1.2 [52].

1.2.2.2 Objective Evaluation Method

The objective evaluation method is based on mathematical comparison between original speech and the synthesized speech. The commonly used measurements may be divided into time domain evaluation and frequency domain evaluation. The time domain evaluation includes SNR, weighted SNR, average segment SNR, and so on. The frequency domain evaluation involves Bark Spectral Distortion (BSD) and MEL spectral measures [52,54].

Table 1.1 Some Speech Coding Standards

Classification	Algorithm	Name	Bit-rate (kb/s)	Standard	Applications	Quality (MOS)
Waveform Coding	PCM	Pulse Code Modulation	64	G.711	Public networks	4.0–4.5
	ADPCM	Adaptive Differential PCM	32	G.721	ISDN	
	SB-ADPCM	Sideband ADPCM	16~40	G.726		
			16~40	G.727		
			48~64	G.722		
Parameter Coding	LPC-10E	Linear Prediction Coding	2.4	FS-1015	Secret communications	2.5–3.5
Waveform-parameter Hybrid Coding	CELP	Code Excited LPC	4.8	FS-1016	Military communications	3.7–4.0
	RPE-LTP	Regular Pulse Excited-Long Term Prediction	13	GSM	Mobile communications	
	LD-CELP	Low Delay-CELP	16	G.728	Public networks ISDN	
	CS-ACELP	Conjugate Structure-Algebraic CELP	8	G.729	VoIP Mobile communications	
	MELP	Mixed Excitation Linear Prediction	2.4	FS-1015	Secret communications	
	MP-MLQ-ACELP	Multipulse-Maximum Likelihood Quantization-ACELP	5.3/6.3	G.723.1	PSTN	
Perceptual Coding	SQVH	Scalar Quantization Vector Huffman Coding	24/32	G.722.1	Public networks	4.2
	MPEG	Multisubband Perceptual Coding	128		CD	5.0
	AC-3	Perceptual Coding (Audio Coding-3)	300		Home theatre	

Table 1.2 MOS Levels and Descriptions

MOS score	Quality level	Degree of distortion
5	Perfect	Like face-to-face conversation or radio reception.
4	Good	Imperfections can be perceived, but still sound clear.
3	Fine	Imperfections can be easily perceived, but sound can be accepted.
2	Bad	Nearly impossible to communicate.
1	Extremely bad	Impossible to communicate.

Table 1.3 Performance Comparisons of Three Speech Coding Methods

Classification	Required bit-rate (kb/s)	Quality (MOS)
Waveform Coding	16–64	4.0–4.5
Parameter Coding	2.4	2.5–3.5
Waveform-Parameter Hybrid Coding	2.4–13	3.7–4.0

Starting from the objective evaluation perspective, Table 1.1 can be redesigned to Table 1.3 [52].

Table 1.3 indicates that waveform coding methods get high MOS scores. They require a higher bit-rate, while the required bandwidth is correspondingly large. Parameter coding and hybrid coding algorithms greatly reduce the transmission bit-rate under the condition that communication requirements can be met.

1.3 **RELATED WORK**

After more than 10 years of research, research on the application of information hiding techniques in the field of communication continues to increase gradually. The application of information hiding techniques in secret communication has been studied by many academic institutions and individuals around the world. Cambridge University started the research on information hiding much earlier than others, and has obtained many remarkable achievements [21,22] on the classification of information hiding techniques, basic theory of steganography, among others. MIT Media Lab attended to the study of how to hide secret messages in images, text, and speech [55]. The Japanese National Defense Academy had proposed a method to embed secret messages in images by using spectrum spread technology [56]. Some famous international companies, for example IBM, also have launched a series of research projects and developed corresponding hardware and software to provide copyright protection services [57–59].

A summary of research achievements of information hiding from an individual perspective includes the following.

Professor Fabien A. P. Petitcolas [60] at the University of Cambridge is one of the pioneers and founders of information hiding. He gives an overview of the field of information hiding first [21], classifies information hiding techniques, and provides studies on steganography techniques.

Professor Pierre Moulin and Joseph A. O'Sullivan [22] at the University of Illinois at Urbana-Champaign performed an information-theoretic analysis of information hiding, forming the theoretical basis for the design of information hiding systems. They first completed the formulation of information hiding as a communication problem, studied on a binary channel with optimal information hiding and attack strategies, and presented optimal information hiding and attack strategies for Gaussian host data.

Professor Naofumi Aoki at Hokkaido University proposed an approach of the lossless steganography technique for telephony communications [61] to mitigate the problem of having no way to avoid inevitable degradation of speech data by embedding secret messages with the LSB replacement technique. This proposed approach exploits the characteristic of the folded binary code employed in several speech codecs, such as G.711 and DVI-ADPCM. He completed an error concealment technique for such degradation by taking account of both sender-based and receiver-based procedure information [62]. In the proposed technique, the sender-based side information, with an improved receiver-based waveform reconstruction technique, was transmitted through steganography, so that its datagram was completely compatible with the conventional format of VoIP. From experimental results, it was indicated that the proposed technique may potentially be useful for improving the speech quality by comparing it with other conventional techniques [62]. He performed a band extension technique for G.711 speech using steganography [63] to investigate a band extension technique for speech data encoded with G.711, the most common codec for digital speech communications systems such as VoIP. The proposed technique employs steganography for the transmission of the side information required for band extension [63]. He conducted a band extension technique for G.711 speech using steganography based on full wave rectification [64]. Due to steganography, these two proposed techniques were able to enhance the speech quality without an increase of the amount of data transmission. They accomplished a technique of lossless steganography for G.711 [65] to realize lossless steganography for G.711 telephony speech [66], which can be used in the most common codecs for digital speech communications such as VoIP.

These two proposed techniques exploited the characteristics of G.711 for embedding steganogram information without any degradation. Aoki achieved lossless steganography techniques for IP telephony speech taking into account the redundancy of folded binary code [67] to investigate the possibility of lossless steganography techniques for G.711 and DVI-ADPCM, the codecs employed in IP telephony services. The proposed techniques exploited the redundancy of the folded binary code to embed steganogram information into speech data without degradation. He presented a method to achieve a lossless steganography technique for G.711 telephony speech by exploiting the characteristic of G.711 for embedding steganogram information into speech data without degradation [68]. This proposed method is a semilossless steganography technique for increasing the capacity of the proposed technique [68].

Aoki fulfilled a lossless steganography technique for DVI-ADPCM transactions on the fundamentals of electronics [69]. The proposed technique was able to embed steganogram information without degradation by exploiting the redundancy of the folded binary code employed in DVI-ADPCM. He presented a semilossless steganography technique for G.711 telephony speech [70] to investigate the possibility of increasing the capacity of the lossless steganography technique. He also carried out enhancement of speech quality in telephony communications by steganography [71]. Two topics, packet loss concealment technique and band extension technique, were explained. These techniques employed steganography for transmitting side information for improving the performance of signal processing. In addition, it described an efficient steganography technique devised for G.711, the most common codec for telephony speech standardized by ITU-T. The proposed technique, named selective LSB replacement, outperformed the conventional one in order to decrease the degradation caused by embedding side information into speech data by steganography [71].

Professor Wojciech Mazurczyk at the Warsaw University of Technology completed the first survey of the existing VoIP steganography methods and their countermeasures [72]. He thought steganographic methods were usually aimed at hiding the very existence of the communication. Due to the rise in popularity of IP telephony, together with the large volume of data and variety of protocols involved, it was currently attracting the attention of the research community as a perfect carrier for steganographic purposes. He proposed a new, lightweight, no-bandwidth-consuming authentication and integrity scheme for VoIP service based on SIP (Session Initiation Protocol) to realize a new VoIP traffic security scheme with digital watermarking [73]. This scheme used a password-share mechanism and adopted digital watermarking to secure the transmitted audio and signaling protocol, the basis of IP telephony. It could greatly improve, if combined with existing security mechanisms, the overall IP telephony system's security. He presented a new security and control protocol for VoIP based on steganography and digital watermarking [74]. The proposed protocol was an alternative to the IETF's (Internet Engineering Task Force) RTCP (Real-Time Control Protocol), which is used for real-time application traffic. Additionally this solution offers authentication and integrity, and it was capable of exchanging and verifying QoS and security parameters. He conducted covert channels in SIP for VoIP signaling to evaluate available steganographic techniques for SIP that could be used for creating covert channels during the signaling phase of VoIP call [75].

Apart from characterizing existing steganographic methods, Mazurczyk provided new insights by introducing new techniques and also estimated the amount of data that could be transferred in signaling messages for a typical IP telephony call [75]. He provided new insights by presenting two new techniques in steganography of VoIP streams [76]. The first one was a network steganography solution that exploited free/unused protocols' fields and was known for IP, UDP, or TCP protocols but had never been applied to RTP (Real-Time Transport Protocol) and RTCP, which were characteristic for VoIP. The second steganographic method, called LACK (Lost Audio Packets

Steganography), provided a hybrid storage-timing covert channel by utilizing delayed audio packets [76]. LACK provided a broader context of network steganography and of VoIP steganography in particular [77]. The analytical results presented in this method concerned the influence of LACK's hidden data insertion procedure on the method's impact on quality of voice transmission and its resistance to steganalysis [77].

Mazurczyk also studied the problem of suspicious VoIP delays [78]. He thought that the modification of an RTP packet stream provided many opportunities for hidden communication since the packets may be delayed, reordered, or intentionally lost. In this study, to enable the detection of steganographic exchanges in VoIP, he examined real RTP traffic traces to figure out the behavior of normal delays in RTP packet streams, and how to detect the use of known RTP steganographic methods based on this knowledge. He achieved hidden communication in IP telephony by using transcoding [79], which presented a new steganographic method for IP telephony called TranSteg (transcoding steganography). Typically, in steganographic communication it was advised for covert data to be compressed in order to limit its size. In TranSteg it was the overt data that was compressed to make space for the steganogram. The main innovation of TranSteg was, for a chosen voice stream, to find a codec that would result in a similar voice quality but smaller voice payload size than that originally selected. Then the voice stream was transcoded. At this step the original voice payload size was intentionally unaltered and the change of the codec was not indicated. Instead, after placing the transcoded voice payload, the remaining free space was filled with hidden data. The TranSteg proof of concept implementation was designed and developed. He conducted many experiments, and results proved that the proposed method was feasible and offered a high steganographic bandwidth. TranSteg detection was difficult when performing inspection in a single network localization [79]. He fulfilled a first practical evaluation of lost audio packets steganographically [80]. This research presented the first experimental results for an IP telephony-based steganographic method called LACK. This method utilized the fact that in typical multimedia communication protocols like RTP, excessively delayed packets were not used for the reconstruction of transmitted data at the receiver; that is, these packets were considered useless and discarded. He performed experiments based on a functional LACK prototype, and results showed the method's impact on the quality of voice transmission. Achievable steganographic bandwidth for the different IP telephony codecs was also calculated [80].

Professor Huang, Y. F. and his team members at Tsinghua University carried out covert communication based on steganography in SIP [81]. They mentioned that an SIP UA (user agent) could dynamically embed some secret message into the speech carrier with good efficiency, and introduced the principle of the secret communication based on steganography technology over VoIP. They proposed a new architecture of the software used to transmit the hidden information. In particular, they adopted a relatively new steganography algorithm, which was referred to as the LSB matching method. In addition, an approach for avoiding the loss of secret data based on Redundant Audio Data was presented [81]. They presented a method of key distribution over the covert communication based on VoIP to exchange the private key by adopting steganography in an RTCP packet [82]. In the proposed method, a model of covert communication based on VoIP is given, and the significance of key exchanging to the covert communication is pointed out. Then, they turned to suggest a steganography algorithm to hide the private key in an RTCP packet, and analyzed the performance of the algorithm. In order to overcome the packet loss, a scheme of key pairs was suggested in case the previous key was lost in transmission [82]. They completed steganography in inactive frames of VoIP streams encoded by the source codec [83]. This study described a novel high-capacity steganographic algorithm for embedding data in the inactive frames of low bit-rate audio streams encoded by G.723.1 source codec, which was used extensively in VoIP, and they proposed a new algorithm for steganography in different speech parameters of the inactive frames. Performance evaluation showed embedding data in various speech parameters led to different levels of concealment. Furthermore, an improved voice activity detection algorithm was suggested for detecting inactive audio frames taking into account packet loss [83].

Huang et al. proposed an approach of detection of covert VoIP communications using sliding window-based steganalysis [84]. This proposed approach utilized a unique sliding window mechanism and an improved regular singular (RS) algorithm for VoIP steganalysis, which detected the presence of LSB embedded VoIP streams. They presented a novel steganalysis method that employed the second detection and regression analyzing steganalysis of compressed speech to detect covert VoIP channels [85]. The proposed method could not only detect the hidden message embedded in a compressed VoIP speech, but also accurately estimated the embedded message length. In order to estimate the hidden message length, this method used a second steganography statistic, which was embedding information in a sampled speech at an embedding rate followed by embedding other information at a different level of data embedding [85].

Professor Wu Zhijun and his research group proposed a research and implementation approach for speech information hiding telephony (SITH) to realize secure communication over PSTN based on the technique of speech information hiding [86]. The SITH system architecture and function module were given in detail, and the overall system was designed in an embedding method based on a digital signal processor (DSP). They completed the design of a speech information hiding telephone, which is an overall system designed for secure and covert communication when speech was transmitted over PSTN [87]. They fulfilled an ABS-based speech information hiding approach to embed dynamic secret speech information data bits into original carrier speech data, with good efficiency in steganography and good quality in output speech [88]. This approach realized speech information hiding based on an ABS algorithm, and adopted a speech synthesizer in a speech coder to implement synchronous speech embedding and coding; that is, the fusion of secret speech information data in speech coding.

Zhijun et al. studied a novel approach of secure communication based on the technique of speech information hiding to establish a Secret Speech Subliminal Channel (SSSC) for speech secure communication over PSTN, and employed the algorithm of ABS speech information extraction to recover the secret information [25].

They proposed an approach of speech information hiding based on the G.729 scheme to implement the ABS-based algorithm of speech information hiding and extraction, which forms the theoretical basis for designing a secure speech communication system [89]. They developed a new steganography scheme based on an information hiding technique for hiding one piece of speech information data in another [90]. In this technique, a host speech as a carrier played a role in two aspects: first, it provided a common communication channel for common information; second, it set up a covertures channel for steganography information (secret message). This technique also verified the SITH system to realize the information hiding technique-based speech secure communication over PSTN [90]. They accomplished the G.711-based adaptive speech information hiding approach [91], which proposed an adaptive LSB algorithm to embed dynamic secret speech information data bits into public speech of G.711-PCM (Pulse Code Modulation) for the purpose of secure communication according to energy distribution, with high efficiency in steganography and good quality in output speech [91].

They also presented an approach for speech information hiding based on the G.721 scheme [92]. The proposed approach achieved ABS-based speech information hiding and extraction algorithms on speech coding scheme G.721 [92]. They achieved an approach of LPC parameters substitution for speech information hiding [24] to realize an information hiding algorithm based on the ABS speech coding scheme. This approach substituted secret speech data bits for linear predictive coefficients in LPC. They performed many experiments, and result statistics showed that the change of the voiced signal track was slow and the codebook for LPC vector was large in the ABS coding scheme. Therefore, a new concept, filter similarity, was proposed to determine the LPC parameters for substituting with secret speech information, and to generate a multicodebook for saving storage space to store secret speech information. To achieve the best effects of information hiding, a dynamic threshold was built to make an optimal trade-off among hiding capacity, security, robustness, transparency, and real time [24].

1.4 ANALYSIS OF AVAILABLE INFORMATION HIDING METHODS

Since the audio data copyright protection and digital security communication system have been more widely used in recent years, studies on speech information hiding are being developed. Many methods that are based on the human auditory system (HAS) have been proposed, and several commonly used approaches related to secure communication are based on the technology of information hiding.

1.4.1 LEAST SIGNIFICANT BIT

The Least Significant Bit (LSB) [93] method is the easiest way to embed secret information. By replacing the minimum weighting value of a sampled speech signal with binary bits of secret information data, the secret information can be hidden in the speech. In the receiver, the only purpose is to extract secret information bits from the corresponding locations. To increase the detection difficulty of secret data, a pseudorandom sequence can be used to control the location, into which the secret binary information is going to be embedded. Although poor in security, the LSB method is simple and easy to implement, embeds and extracts information fast, and has a high hiding capacity [24,94,95].

1.4.2 PHASE HIDING METHOD

The human ear's hearing system is not sensitive to the absolute phase of the sound. This finding is also applied in speech information hiding. In this phase hiding method, the absolute phases of speech information should be replaced by the reference phases, which represent the secret information. To ensure the fixed relative phase between the signals, all subsequent signals change the absolute phase at the same time. In the receiving end, phase detection is done according to the synchronization mechanism [96,97]. To improve the accuracy of the information extraction, generally phase offset can be set at $\pi/2$. In general, the hiding capacity of the phase hiding method is 8 to 32 bits per second [98]. The more noise the speech signal itself contains, the greater relative channel capacity this method achieves. Compared with the LSB algorithm, the phase hiding method hides a smaller amount of data, but this method has significantly improved the ability to fight against noise attacks in an antiattack.

1.4.3 ECHO HIDING METHOD

According to the human ear characteristic of audio signals, if a weak signal appears after the strong signal in a very short period of time (typically 0–200 milliseconds), then the weak signal will be unhearable. Utilizing this hearing sense characteristic, the speech echo hiding method tries to introduce the echo signal $f(t-\Delta t)$ in a time-discrete signal $f(t)$, gets the disguised signal $c(t) = f(t) + \lambda f(t-\Delta t)$, and then the purpose of hiding can be achieved. An embedding schematic diagram of the speech echo hiding method is shown in Figure 1.2 [99].

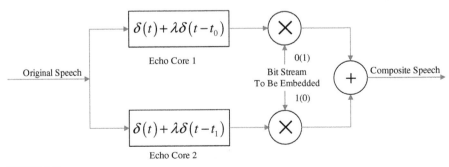

FIGURE 1.2

Schematic diagram of speech echo hiding.

Two information parameters can be hidden using the echo hiding method. The receiver computes the autocorrelation function of speech cepstrum under the synchronization information [100,101]. According to the different peak locations and magnitude of the autocorrelation function, secret information could be obtained. Due to the human ear characteristics, the range of delay time (t_0 or t_1) is generally set as 50 to 200 ms and λ may be around 0.7 to 0.8 ms [99].

The speech echo hiding method has good transparency. Usually it is still able to restore the hidden information after being attacked.

1.4.4 HIDING METHOD BASED ON STATISTICS

The basic idea of this approach is to embed secret information into a certain statistic value of audio data, to hide information by modifying a certain statistical value [102]. Taking the general signal processing and attacks into account, the cepstrum parameters alter a little with regard to that in time-domain sampling. An information hiding method based on cepstrum forcibly modifies the average value of the cepstrum parameter under the secret data 0 or 1. That should be done after obtaining the cepstrum of speech signal. Then it recovers the speech and encodes. Thus the secret information is embedded into the statistic average value of the chosen cepstrum parameter. The process of cepstrum parameters calculation is shown in Figure 1.3 [102]. In the embedding procedure, an intuitive psycho-acoustic model is used for controlling the distortion degree of introduced signals. In the receiving end, according to the synchronization mechanism, the receiver can judge what has been embedded by computing cepstrum parameters of received speech signals.

Experimental results indicate that this new hiding method can reach a hiding capacity of 20 bits per second, and subjective hearing tests also show that the quality of composite speech generally equals the 64 bits per second encoded speech signal quality, which can be widely accepted by people. It is enough to see the transparency and robustness. Compared with the classic SS (spread spectrum) technology, when you cannot get the original data, especially the synchronization structure of audio data in the detection process, the hiding technology based on statistics tends to be more robust, but the hiding capacity appears to have some limitations.

1.4.5 TRANSFORM DOMAIN METHOD

The transform domain method has been widely used in watermarking technology, and has become more and more popular in audio communications. The basic idea is that it embeds secret information in a certain transform domain of digital works

FIGURE 1.3

The process of cepstrum parameters calculation.

so that secret information can be hidden in the most important part of the carrier [103–105]. Thus, as long as the attackers do not excessively destroy the carrier, the hidden information can be preserved.

Commonly used transform domain methods are Discrete Fourier Transform (DFT), Discrete Cosine Transform (DCT), Hadarma Transform, wavelet transform, and Modulated Complex Lapped Transform (MCLT). These methods embed secret information into the coefficients of the frequency domain, and utilize a spread spectrum technique for effective coding. Transparency and robustness are improved, and they also use filtering technology to eliminate high frequency noise included by the hiding process. Thus the transform domain methods become more powerful to fight against low frequency filtering attacks. The principle of the transform domain method is shown in Figure 1.4 [104].

The experimental results show that when the transform domain method is used for speech information hiding, it can better resist the attacks of signal processing and keep the insensibility for human hearing. These methods have become a hot topic in the speech information hiding field.

Based on the analysis of the existing methods, results show that they have certain overall disadvantages for hiding capacity, robustness, and speech quality, although the existing methods have certain characteristics individually.

Secure communication requirements of real-time, large amounts of data, and fast data rate are challenges to the approach of secure communication based on speech information hiding. To achieve the best balance between hiding capacity, speech quality, and robustness, a reasonable solution is to start from the speech coding standard.

There is a lot of redundant information in speech signals, which is the premise for realizing low bit-rate speech coding. The purpose of speech coding is to eliminate this redundancy maximum.

All the parameters are transmitted frame by frame in all the traditional LPC-based ABS coding methods. The disadvantages of this approach are two-fold [24,88]:

- The LPC-based ABS coding method does not take into account the characteristic that voiced speech signals change slowly in the vocal tract during the production process. This characteristic means that a lot of redundant

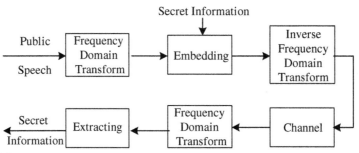

FIGURE 1.4

The principle of the transform domain method.

information exists in the process of voiced signal generation. Therefore, from the viewpoint of encoding, this characteristic can be used to remove redundancy further and reduce the code rate. Otherwise, from the perspective of information hiding, it is how to make use of this characteristic in hiding secret speech information.

- The LPC-based ABS coding method adopts the vector quantization method to represent the LPC coefficients of the synthesis filter. This quantization method can significantly reduce the number of bits representing the LPC coefficients. However, in the traditional method, a single codebook is obtained by the overall design of the LPC vectors in all speech samples. This design idea does not take into account the characteristics of different speech segments, resulting in limited efficiency of quantization, and the codebook is relatively large, which needs large storage space and a larger search amount.

Therefore, these two characteristics are used to design the proposed algorithm of LPC parameters substitution for secret speech information hiding based on filter similarity (LPC-IH-FS) and information extraction algorithm with blind detection-based minimum mean square error (MES-BD). There are two issues for embedding secret speech in public speech and communication over a network to realize secure communication:

- Figuring out the similarity between neighboring speech frames by using the characteristic that the vocal tract of voiced speech signals changes slowly.
- Classifying the LPC coefficients by adopting the different characteristics of each speech segment, and generating the different vector quantization codebook by quantizing different types of LPC coefficients. This work will help to improve the quantization efficiency by reducing storage space and search volume.

1.5 ORGANIZATION OF THIS BOOK

In this book, a number of methods to hide secret speech information in accordance with different types of digital speech coding standards are described. During the past 10 years, the continued advancement and exponential increase of network processing power have enhanced the efficacy and scope of speech communication over networks. Therefore, the author summarized years of research achievements in the speech information hiding realm to create this book, containing a mathematical model for information hiding, the scheme of speech secure communication, the ABS-based information hiding algorithm, and the implemented speech secure communication system. These topics are organized into sections in accordance with the security situation of networks, and include nine chapters that introduce speech information hiding algorithms and techniques (embedding and extracting) that are capable of withstanding the evolved form of attacks.

Each chapter begins with a high-level summary for those who wish to understand the concepts without wading through technical explanations, followed by concrete examples and more details for those who want to write their own programs. The combination of practicality and theory allows engineers and system designers not only to implement the true secure communication procedures, but also to consider probable future developments in their designs.

Chapter 1 gives a general introduction to the background of secure communication based on information hiding, the application of information hiding, and speech coding technology. Basic definitions and related concepts in the realm of speech information hiding are introduced. The fundamentals will help readers comprehend the new technology of information hiding. To help readers better understand the development of the research contents of this book, the related works are analyzed to show other researcher's achievements. A comparative analysis of the existing information hiding methods is given, and the principle and the working process for the five kinds of existing methods are summarized. Finally, the organization of this book is introduced.

Chapter 2 presents a speech information hiding model for transmitting secret speech through subliminal channels covertly for secure communication over PSTN or VoIP. The proposed model for secure communication based on the technique of information hiding has more severe requirements on the performances of data embedding than watermarking and fingerprinting in the aspects of real time, hiding capacity, and speech quality. The main theme is that the embedding operation of a secure communication system has a lot of uncertainty for the attackers. The security analysis of this model by means of information theory and actual testing proved that it is theoretically and practically secure.

Chapter 3 presents information hiding and extraction algorithms based on an ABS speech coding scheme for the purpose of real-time speech secure communication. This hiding algorithm substitutes secret speech data bits for linear predictive coefficients (LPCs) in linear predictive coding (LPC). Statistics show that the change of voiced signal tracks is very slow and the codebook for the LPC vector is very large in the ABS coding scheme. Therefore, a new concept, filter similarity, is proposed to determine the LPC parameters, replacing them with secret speech information, and to generate a multicodebook for saving storage space for secret speech information. Then, the proposed algorithms of LPC parameters substitution for secret speech information hiding based on filter similarity (LPC-IH-FS) and secret speech information extraction with blind detection-based minimum mean square error (BD-IE-MES) are introduced in detail. To achieve the best effects of information hiding, a dynamic threshold is set up to make an optimal tradeoff among hiding capacity, security, robustness, transparence, and real time.

Chapter 4 proposes a novel approach for hiding 2.4 kbps MELP (Mixed Excitation Linear Prediction) coded secret speech information in 32 kbps G.721-ADPCM (Adaptive Differential Pulse Code Modulation) for real-time secure communication. This approach adopts the ABS-based algorithm to embed, transmit, and extract secret speech. G.721 is a technique of waveform coding, which compresses data by adopting the high correlation between samples and adaptive step-size. It is a popular speech coding scheme widely used in stream media communication and telecommunication, such as Digital Enhanced Cordless Telecommunications (DECT), which is specified up to bit-accurate test sequences provided by the International Telecommunication Union (ITU), therefore differences in functionality can be easily checked and removed.

Chapter 5 introduces an ABS-based approach of hiding 2.4 kbps MELP coded secret speech information in G.728 Low Delay-Code Excited Liner Prediction (LD-CELP) for secure communication. Similar to the previous chapters, the ABS idea is adopted. First, the CELP production model is illustrated and the coding operation is introduced. Then embedding and extraction algorithms are carried out with experiments and analysis.

Chapter 6 is devoted to the G.729 CS-ACELP coder. The most distinguishing features are using the algebraic excitation codebook and conjugating VQ for the involved gains. The chapter starts with the evaluation of the structure of the algebraic codebook, followed by the construction and search methodology of the adaptive codebook. The description of encoding and decoding operations, algebraic codebook search techniques, and gain quantization are given.

Chapter 7 proposes a novel approach of secure communication based on the technique of speech information hiding. The proposed approach adopts the GSM speech coding scheme based on Regular Pulse Excited-Long Term Prediction (RPE-LTP) to establish a secret speech subliminal channel for speech secure communication over PSTN and VoIP by using the proposed ABS speech information hiding LPC-IH-FS algorithm and extracting the BD-IE-MES algorithm. This chapter combines the ABS concept with the RPE-LTP coding scheme to illustrate speech information hiding. A simple coding scheme is given and the detailed embedding and extraction operations are carried out. Experimental results show that this approach is reliable, covert, and securable.

Chapter 8 makes use of the most widely used voice communications system—VoIP—which is the fundamental communication IP protocol. In this chapter, a model for VoIP-based covert communications is put forward based on the analysis of the characteristics of VoIP, and an approach of embedding secret speech information in VoIP G.729 speech flows is proposed based on matrix coding. The implementation of embedding and extraction are illustrated.

Chapter 9 introduces the design and implementation of a real-time speech secure communication over PSTN, and shows how to implement the secure communication over PSTN by using the information-hiding-based model and algorithm. The scheme of real-time speech secure communication over PSTN is introduced. The detailed design of Speech Information Hiding Telephony (SIHT) using speech information hiding technology is presented.

The Information Hiding Model for Speech Secure Communication

The theory of information hiding is a new information security theory developed on the basis of Shannon's great works, *A Mathematical Theory of Communication* [22,106] and *Communication Theory of Secrecy Systems* [21,107]. As Pierre Moulin and Joseph A. O'Sullivan [22,108] mentioned, information hiding embeds secret information in a host data set and is reliably communicated to a receiver, which clearly indicates that information hiding belongs to the communication field. The information hiding theory [22,108] provides the theoretical basis for the covert communication of secret information. Therefore, the communication model based on the technique of information hiding is necessary for speech secure communication.

2.1 INTRODUCTION AND MOTIVATION

Information hiding (i.e., data embedding) is a communication issue [22] with two important parts: signal sources and communication channels. The concept of information hiding is to conceal an important secret message in public information. A survey of current information hiding according to applications is given by Petitcolas et al. [21], who classified information hiding into several research areas such as watermarking, fingerprinting, and steganography. Watermarking and fingerprinting are applied for authentication and authority, which relate to signal sources of communications for the purpose of copyright protection of digital media. Steganography is the art of covered or hidden writing, which is covert communication for secret messages to conceal the existence of a message from malicious attackers eavesdropping on a communication channel. Nowadays, most formal models [108–110] of information address the problem of copyright protection for digital media; that is, these models focus on the protection of signal source of communications for designing steganographic systems to generate watermarking and fingerprints. The application of watermarking in image, audio, or video data is most popular. Meanwhile, little research attention has been paid to the security of the covert communication channel. Currently there is no universal information hiding model available for protecting the secure communication channel.

Available information hiding models are used mainly for copyright protection schemes. These models are different from the one considered here for secure communication. From a communication point of view, the model used for copyright protection is static; the operations of watermarking embedding and extracting are performed independently, and there are no real-time requirements between them. Once the watermarking is embedded into the protected media at the sender end and transmitted to the receiver end, the watermarking can be extracted at any time for the purpose of verification. On the contrary, the model used for secure communication is dynamic—the operations of secret speech information embedding and extracting are completed at the same time, and require both sides of secret information communication in real time. Hence, the work principle of secure communication based on the information hiding technique is the same as confidential telephony, which is real-time secure speech communication. This is the focus of this book. The model based on the technique of information hiding has more demanding requirements on the performance of data embedding than watermarking and fingerprinting in such aspects as real time, hiding capacity, and speech quality. Under the same constraints of speech quality and security, secure communication needs larger hiding capacity and a higher embedding rate than watermarking and fingerprinting. For the purpose of secure communication, this chapter proposes a novel and practical speech information hiding model for real-time speech secure communication over PSTN or VoIP.

This chapter focuses on the steganography of information hiding, called covert communication, which expands the application of information hiding to the area of real-time speech secure communication.

For convenience, the notations and abbreviations used throughout the book are summarized in Tables 2.1 and 2.2 respectively [22].

Table 2.1   Notations Used in the Book

Symbol	Definition	Abbreviation	Definition
X	Original public speech	Emb	Embedding function
X'	Carrier speech	Ext	Extract function
X''	Composite speech	$Q(y\|x)$	Attack channel
$\hat{X}$	Extracted public speech	$C(M'',R,S)$	Key subliminal channel
$\hat{X}'$	First intermiddle composite speech	m	Message digest, $m \in M$
$\hat{X}''$	Second intermiddle composite speech	R	Signature parameter, $r \in R$
M	Original secret speech	S	Signature parameter, $s \in S$
$\hat{M}$	Extracted secret speech	C_H	Hiding capacity
M'	Secret speech in MELP 2.4 kbps	D	Degradation of speech quality
M''	Secret speech	$p(x)$	Probability Mass Function (PMF)
K	Private key	H	Difficult for crack
$E(M)$	Encryption	I	Entropy
$D(\hat{M})$	Decryption	Y	Attacked speech

Table 2.2   Abbreviations Used in the Book

Abbreviation	Definition	Abbreviation	Definition
ABS	Analysis-by-Synthesis	ABS-SIH	Analysis-by-Synthesis Speech Information Hiding
LPC	Linear Predictive Coding	SSSC	Secret Speech Subliminal Channel
LSF	Linear Spectrum Frequency		
KSC	Key Subliminal Channel	CELP	Code Excited Linear Prediction
ElGamal	Signature scheme	MELP	Mixed Excitation Linear Prediction

The rest of this chapter is organized as follows: Section 2.2 introduces the model of information hiding as a communication problem, and gives the general model for information hiding application. Section 2.3 proposes the model of speech information hiding for secure communication, illustrates the key elements used in this model, and analyzes the performance of the proposed model on hiding capacity, security, and speech quality. Section 2.4 completes the model performance test, and shows the analysis result. Section 2.5 concludes this chapter with a summary.

2.2 MODEL OF INFORMATION HIDING AS A COMMUNICATION PROBLEM

The purpose of information hiding is transmitting secret information to the destination covertly through camouflage and deception. A schematic diagram of information hiding as a communication problem is shown in Figure 2.1 [22,111].

Figure 2.1 is a simplified block diagram of a practical communication system in which the encoder and decoder play the roles of embedding and extracting in the information hiding system. Secret information is hidden in the encoder at the sender end, and it is extracted from the decoder at the receiver end. Secret information is safely transmitted from the sender to the receiver via a communication channel.

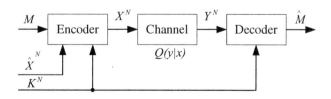

FIGURE 2.1

Schematic diagram of information hiding as a communication problem.

In Figure 2.1, M is a secret message, $\hat{X}^N$ is the host data set, and X^N is the composite data set, which is generated by embedding M into $\hat{X}^N$ in the encoder. The composite data set X^N is transmitted though the communication channel under attack $Q(y|x)$. Y^N is the composite data set produced when X^N is experiencing the attack $Q(y|x)$, and then Y^N is the input to the decoder. $\hat{M}$ is the recovery secret message that was generated in the decoder. K^N is a shared key of both the encoder and decoder.

Commonly, the technology of information hiding is embedding secret information M into host data set X^N to generate a new host data set X, which is transmitted to the receiver through the communication channel. During the transmission, X is analyzed by data processing operations, which try to extract information about M and remove it. In fact, the operation of data processing belongs to the attack on X. Therefore, the information hiding system must meet two basic requirements as follows [108,109,112].

Information hiding has two important criteria. The first is transparency or unobtrusiveness, which requires the data set $\tilde{X}$ to be very close to X. This requirement is performed in accordance with an appropriate test standard. The second criterion is robustness, which requires that no matter what level of data processing is performed on X, such as transform in time or frequency domain and filtering or amplitude limiting, the secret information M must be kept hidden. Usually, there is a certain limited amount of distortion that is intentionally introduced against the attack.

The requirements of information hiding are mainly dependent on the hiding and extraction algorithm performance. In order to achieve the two requirements mentioned previously, hiding and extraction modules are two key issues to consider in the model design.

The general model for information hiding application is shown in Figure 2.2 [22,113,114].

The terms used in Figure 2.2 are described as follows:

- **Public Data X (or original information)** refers to the data that has not been embedded with secret data (or secret information). It is the carrier for secret data and it may be plain text or original data.
- **Secret Data M (or secret information)** is the data that is going to be embedded into original data and that needs to be secretly transmitted.

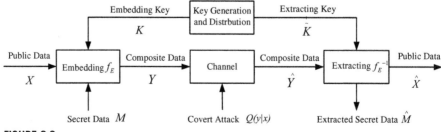

FIGURE 2.2

General model for information hiding application.

- **Embedding Key K** is the key used in the procedure of embedding secret data into original data.
- **Embedding Algorithm** is an operation that uses a mathematical function f_E to embed secret data into original data. This operation involves the embedding key K at the same time.
- **Composite Data Y and $\hat{Y}$** refer to the data that are generated by embedding secret data in the original data. Y is transmitted through the channel under covert attack $Q(y|x)$, and then composite data Y is generated. Y is the information that is really transmitted.
- **Extracting Key $\tilde{K}$** is the key used in the procedure of extracting secret data from composite data.
- **Extracting Algorithm f_E^{-1}**, in relation to the embedding algorithm, uses a mathematical function f_E^{-1}, which extracts secret data from composite data. This operation of secret data extraction is completed in cooperation with the extracting key K.
- **Covert Attack $Q(y|x)$** usually refers to the decryption operation or computation, which aims at digging out secret data. A covert attack can be divided into two categories: positive attack and passive attack. In terms of a cryptographic attack, a covert attack includes several attack methods, such as Cipher text only, known as plain text, chosen plain text, chosen Cipher text, and chosen text.
- **Extracted Secret Data $\hat{M}$** is the secret data received by the receiver. The receiver processes the composite data by using the extracting function f_E^{-1}, and then extracted secret data may be obtained.

Figure 2.2 improves on Figure 2.1, replacing the encoder with embedding f_E and the decoder with extracting f_E^{-1}. f_E and f_E^{-1} represent the embedding and extraction algorithms respectively. The secret data M is embedded in public data X when embedding f_E to generate composite data Y in the action of embedding K. The composite data Y suffers from the covert attack $Q(y|x)$ being transmitted through the channel. The composite data Y is changed into $\hat{Y}$ after transmission in the channel under attack $Q(y|x)$. The secret data $\hat{M}$ is extracted from $\hat{Y}$ in extracting f_E^{-1} under the action of extracting $\tilde{K}$, while the public data $\hat{X}$ is the output from extracting f_E^{-1}.

2.3 SPEECH INFORMATION HIDING MODEL

The proposed model of speech information hiding for secure communication is shown in Figure 2.3. In the following analysis, random variables are denoted by lower case letters (e.g., m), and they take values on the set denoted by upper case letters (e.g., M). All the terms can be referenced to Pierre Moulin's paper [22,114].

Before the procedures of embedding, the preprocessing of original public speech X and original secret speech M includes: (1) M coded in 2.4 kbps Mixed Excitation Linear Prediction (MELP) obtained M' and encrypted by chaos sequence

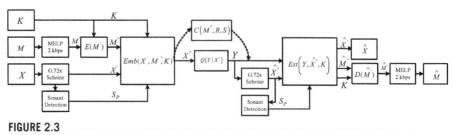

FIGURE 2.3

The model of information hiding.

cipher $E(M)$, which generates the secret speech M''; and (2) the X coded in the G.72x scheme, which generates the carrier speech X'. Both X and M are Pulse Coding Modulation (PCM) speech at the rate of 64 kbps.

Input X' represents the untreated carrier data, and M'' will be embedded into X' following the embedding function $Emb(X', M'', K)$. The resulting data, called composite speech X'', contains the message M''. X'' is a meaningful, continuous, and understandable segment of speech, which is disturbed by malicious attackers (active) and eavesdroppers (passive) when being transmitted through the attack channel $Q(Y \mid X'')$, which then outputs the disturbed composite speech Y. $Q(Y \mid X'')$ is the public channel under attack, in which an SSSC [114–116] is established by the proposed algorithm of Analysis-by-Synthesis Speech Information Hiding (ABS-SIH) [88,114] for secret speech secure communication. Y is the input signal at the receiver end. It is split into two identical signals. One is input to the extracting function $Ext\left(Y, \hat{X}'', K\right)$ as a reference signal, and the other is decoded in a G.72x scheme, which generates the first intermiddle composite speech $\hat{X}''$ to be extracted. The key K is the seed for the chaos sequence cipher, transmitted through the Key Subliminal Channel (KSC) $C(M'', R, S)$ [115,116], which is established by the *ElGamal* [115,116] signature scheme. Here, M is the set of all secret message digest m, $m \in M$; R is the set of signature parameter r for M, $r \in R$; and S is the set of practical signature parameter s for M, $s \in S$. The operation $Ext\left(Y, \hat{X}'', K\right)$ extracts the embedded data to $\hat{M}''$ and outputs the second intermiddle value of composite speech $\hat{X}'$. Then, $\hat{X}'$ is processed to extracted public speech $\hat{X}$. $\hat{M}''$ is decrypted in $D(\hat{M}')$ and with private key K obtained the intermiddle value of secret speech $\hat{M}'$, which is decoded in the MELP 2.4 kbps scheme generating the extracted secret speech $\hat{M}$. Theoretically, $\hat{M}$ should be equal to M and in most cases $\hat{X}$ is the same as X. From the viewpoint of secure communication though, $\hat{X}$ is not of much interest for attackers and eavesdroppers.

In the proposed model, in order to ensure that the periodic operation of embedding and extraction is done in precise positions, the speech sonant period S_P should be determined through the operation of sonant detection.

The design of a speech information hiding model for secure communication focuses on security, hiding capacity, and speech quality. In the proposed model shown in Figure 2.3, information hiding adopts a blind detection technique, and no side information is available at all.

2.3.1 HIDING CAPACITY

According to Pierre Moulin's theory [22], information hiding can be treated as a communication problem. Hence, based on Moulin's research achievements [22], the hiding capacity of the speech information hiding model is determined by the following definition [114].

> **Definition.** Let (Emb, Ext, M, K) be a speech information hiding process subject to degradation of speech quality D, and the Probability Mass Function (PMF) of x' be $p(x')$; the hiding capacity of this process is [21,22,111,113,114]

$$C_H(Emb, D) = \sum_{x'} p(x')r(Emb, x'),$$

where $r(Emb, x') = \log|M(x')|$.

In practical applications, according to the specific use of the speech coding scheme and the minimum MSE rule, the actual maximum hiding capacity C_H is closely related to a few factors, for example the degree of speech quality degradation.

2.3.2 SECURITY

The security evaluation of this model is executed from two aspects: practical utility and theoretical analysis. Similar work using information theory to evaluate the security of secure communication systems was done by J. Zöllner et al. [108], which is very suitable for evaluating the proposed model in this paper, and the analysis result is compatible with this model.

Assume that the attackers or the eavesdroppers have the following conditions [108]:

- The awareness of secret speech embedded into a composite speech
- The knowledge of the embedding and extraction algorithm
- The skill and ability to perform an attack on the system of secure communication
- Unlimited time and resources.

If the attackers or the eavesdroppers try their best without being able to prove and confirm the hypothesis that the existing secret message is hidden, this system is said to be "information theoretically secure."

Suppose that even the attacker or eavesdropper knows clearly that a secure communication in a public network exists, but they have no knowledge of [108]

- The speech coding scheme of the secret speech
- The encryption algorithm for the secret speech
- The embedding and extraction algorithm.

If the attacker or eavesdropper cannot find the channel for secure communication by detection and analysis of secret speech, this system is said to be "application practically secure."

This model has the feature of multilevel security of secret speech protection. First, it is difficult for eavesdroppers to decode the secret speech without knowledge of the secret speech coding scheme. Second, it is difficult for the eavesdroppers to crack the encryption algorithm and decrypt the message of secret speech. Third, it is impossible for eavesdroppers to monitor all G.72x-based communications in the network because G.72x-based stream media communication is widely applied for Internet and telecommunications. Therefore, this proposed approach can provide high security for secure speech communication.

Secret speech is transmitted through SSSC [114–116]. The previous analysis shows that the security of secure communication is closely correlative to:

- The number of transmit line n
- The average call number C_j
- The holding time (average talking time) T_h
- The interval time between calls T_{Inter}
- The Erl in unite time α
- The coefficient of random call λn is determined by the Erlang **B** distribution [117]:

$$En(\alpha) = \frac{\alpha E_{n-1}(\alpha)}{n + \alpha E_{n-1}(\alpha)},$$

where $\alpha = \lambda T_h$.

It is presumed that the attacker or the eavesdropper knows all the information about secure communication, such as the coding scheme of secret speech and public speech, encryption algorithm, and embedding and extraction algorithms. If the attacker or eavesdropper wants to determine the transmit line for secure communication, he or she has to:

- Monitor all calls
- Analyze the traffic of every call
- Check the data in G.72x coding
- Decrypt the secret speech data
- Decode secret speech data in MELP coding.

Each of these jobs needs a certain amount of time to finish. The accurate time for an attacker or eavesdropper to determine the transmit line can be obtained by means of statistical tests.

T_d is the time that an attacker or eavesdropper spent on the detection and analysis of secret speech hidden in PSTN or VoIP. Therefore, if $T_d > T_h + T_{Inter}$, it is concluded that the current secure communication is secure in practice.

The proposed model adopts the *ElGamal* [115,116] signature scheme to transmit secret keys through KSC for the chaos sequence cipher [116]. The embed process of subliminal information E in the *ElGamal* [115,116] digital signature scheme is accomplished with the signature process. The security of a signature algorithm is based on the assumption that it is difficult to solve discrete logarithms in a finite field. Under this assumption, $H(K) = H(K/(M,R,S))$, that is, $I(K;(M,R,S)) = 0$. The uncertainty of subliminal information E to a public receiver cannot be decreased by knowing the triple (M,R,S) of the signature. Hence the conclusion, $H(E) = H(E/(M,R,S)) = H(K/(M,R,S))$, which expresses that the difficulty of obtaining the subliminal information E from the triple (M,R,S) of the signature is equal to attacking the key k [115,116].

2.3.3 SPEECH QUALITY

The evaluation of speech quality focuses on quality changes between:

- Carrier speech X' and composite speech X
- The original secret speech M and extracted secret speech $\hat{M}$.

In the model, the embedding algorithm adopts the ABS-based embedding and extraction algorithm [88]. The operation of embedding is executed under the control of the Minimum Square Error (MES) rule, which limits the variation of speech quality within 3 dB (i.e., $D_1 \leq 3$ dB).

The indicators for speech quality evaluation are uniform correlation coefficient ρ, segment average SNR (signal-to-noise ratio), speech energy variation δ, and segment average *Itakura* distance D_M [39,41]. ρ, SNR, and δ are used for detecting the difference between carrier speech X' and composite speech X. D_M is used for detecting the difference between original secret speech $\hat{M}$ and extracted secret speech $\hat{M}$ [118].

2.3.3.1 Uniform correlation coefficient ρ

ρ expresses the correlation between carrier speech X' and composite speech X [118,119]:

$$\rho = \rho(X,X') = \frac{\sum_i x(i)x'(i)}{\sqrt{\sum_i x^2(i)}\sqrt{\sum_i x'^2(i)}}.$$ (2.1)

The higher the ρ value is, the better the quality of composite speech X. Higher ρ means that the similarity between carrier speech X' and composite speech X in waveform is higher.

2.3.3.2 Segment average SNR

Segment average *SNR* is defined as the average value of *SNR* in each segment speech used for the quality evaluation of carrier speech X' and composite speech X. If $x(k) = 0$, $SNR = 100$.

Define *SNR* as

$$SNR = \frac{1}{P}\sum_{k=0}^{P-1} SNR_k \text{ [119]},$$

where

$$SNR_k = \begin{cases} 100, & x(k)=0 \\ \sum_{i=0}^{L-1} \dfrac{x^2(k \times L+i)}{[x(k \times L+i)-x'(k \times L+i)]^2}, & x(k)=1 \end{cases} \quad 0 \leq k < P.$$

Higher *SNR* means that carrier speech X' and composite speech X have a similar speech quality.

2.3.3.3 Speech energy variation δ

The change of speech quality is measured by the speech quality variation δ, which is calculated according to the equation

$$\delta = E_{no} - E_{nc},$$

where E_{no} is the segment average energy of the original carrier speech X', and E_{nc} is the segment average energy of composite speech X.

Smaller δ means that the speech quality difference between carrier speech X' and composite speech X is small.

2.3.3.4 Segment average Itakura distance D_M

The segment average *Itakura* distance D_M is adopted to measure the difference between spectra of secret speech in the frequency domain [39,41].

Given A_k is a feature vector constructed by the LPC coefficient a_i of order Q, which belongs to the k th frame of original secret speech $M(k)$; that is, $A_k = \left[a_0, a_1, \cdots, a_Q \right]$. B_k is a feature vector constructed by the LPC coefficient b_i of order Q, which belongs to the k th frame of the extracted secret speech $\hat{M}(k)$; that is, $B_k = \left[b_0, b_1, \cdots, b_Q \right]$.

R_k is the $(Q + 1) \times (Q + 1)$ order autocorrelation matrix of $M(k)$:

$$R_k = \begin{pmatrix} r_k(0) & r_k(1) & r_k(2) & \cdots & r_k(Q) \\ r_k(1) & r_k(0) & r_k(1) & \cdots & r_k(Q-1) \\ r_k(2) & r_k(1) & r_k(0) & \cdots & r_k(Q-2) \\ \vdots & \vdots & \vdots & \vdots & \vdots \\ r_k(Q) & r_k(Q-1) & r_k(Q-2) & \cdots & r_k(0) \end{pmatrix}, \tag{2.2}$$

where $M_k(n)$ is the n th sample value of the k th frame; in a practical experiment, $Q = 10$.

The distortion in the frequency spectrum between original secret speech $M(k)$ and extracted secret speech $\hat{M}(k)$ is measured by *Itakura* distance [120]:

$$d_k(A_k, B_k) = \ln \left\{ \frac{A_k R_k B_k^T}{A_k R_k A_k^T} \right\},$$

where T means the transpose of a matrix.

The segment average *Itakura* distance D_M is obtained [120]:

$$D_M = \frac{1}{N_{Frame}} \sum_{k=0}^{N_{Frame}-1} d_k(A_k, B_k),$$

where N_{Frame} is the frame number of secret speech.

Smaller D_M means that the speech quality change between original secret speech M and extracted secret speech $\hat{M}$ is small.

Generally, smaller δ and D_M and bigger *SNR* and ρ mean better speech quality.

2.4 EXPERIMENTS AND RESULTS ANALYSIS

Based on the proposed model, the system for secure communication is obtained to transmit secret information covertly whenever the distribution of composite speech is close to the carrier speech for an attacker or eavesdropper with no knowledge of the speech coding scheme, encryption algorithm, secret key, and embed algorithm.

Experiments based on the speech information hiding model are conducted for the purpose of testing the performance of secure communication systems over PSTN in three aspects: hiding capacity, speech quality, and security.

2.4.1 HIDING CAPACITY

In the implementation of the proposed model to practical applications, experiments on the hiding capacity by embedding secret speech coded in the 2.4 kbps MELP scheme into public speech coded in four G.72x schemes are conducted. Test results (see Table 2.3) show that this model obtains a higher hiding capacity, which meets the requirements for secure communication over PSTN or VoIP.

2.4.2 SECURITY

Experiments on the detection of a secure communication over PSTN or VoIP have been done within the range of one selected switcher with 500 telephone numbers or calls. The following statistical data is assumed:

- Five hundred PSTN telephones or VoIP calls, Erlang B distribution $En(\alpha) = 0.001$
- Holding time (average talking time) is $T_h = 3027$ seconds
- Average call number $C_j = 31$ in one hour

The Erl in unit time can be calculated as $\alpha = \dfrac{3027}{60 \times 60} = 0.841$ Erl. To calculate $En(\alpha)$ by inputting α into the equation $En(\alpha) = \dfrac{\alpha E_{n-1}(\alpha)}{n + \alpha E_{n-1}(\alpha)}$, and then through searching the Erlang B table, the number of transmit lines $n = 8$ is obtained.

Statistical data from the tests show the consuming time for speech coding and encryption in secure communication as the following. (Note: These data are obtained from a PC, not a special hardware processor.)

- Time for G.729 decoding is $T_{G729} = 256$ ms

Table 2.3 Hiding Capacity

Scheme	Hiding capacity (bps)
G.721	1600–3600
GSM	Maximum 2600
G.728	1600–3200
G.729	Maximum 800

- Time for MELP decoding is $T_{\text{MELP}} = 240$ ms
- Time for chaos sequence cipher is $T_{\text{Chaos}} = 374$ ms.

It is reasonable to suppose that every call for secure communication takes an average time of $T_h = 180$ s. The interval T_{Inter} between calls is 240 s. If the time T_d, which is the time the attacker or the eavesdropper spends on detection of secure communication, is longer than 420 s (i.e., $T_d > T_h + T_{\text{Inter}}$), it means that the secure communication is secure in practice.

The time T_d is calculated as

$$T_d = \sum_{i=1}^{n} C_n^i \times (T_{\text{G729}} + T_{\text{MELP}} + T_{\text{Chaos}}) = 1120.0 \text{ s}.$$

Results show T_d is 1120.0 s, which is larger than the summation of T_h and T_{Inter} (i.e., $T_d > T_h + T_{\text{inter}}$).

The evaluation value of the security of this model is

$$S = 1 - \frac{T_h + T_{\text{Inter}}}{T_d} = 1 - 0.375 = 0.625 = 62.5\%.$$

If the number of PSTN telephones or VoIP calls increases, then the number of transmit lines n increases. Test results (see Table 2.4) show the evaluation value of the security of secure communication in a different number of transmit lines n.

With the increase of n, T_d is too big to be accepted by the attacker or the eavesdropper, and the security S of secure communication is increased exponentially. This indicates that the proposed model is information-theoretically and application-practically secure.

2.4.3 SPEECH QUALITY

In the test of the proposed model, experiments are made by adopting the three different embed algorithms, such as ABS [88], adaptive LSB [121], and LSB [99], to test the performance of the information hiding model. Test results (Table 2.5) [114] show

Table 2.4   Test Results for a Different Number of Transmit Lines n

Number	$n = 8$	$n = 16$	$n = 24$	$n = 32$
Security	62.5%	75.%	82.5%	95.%

Table 2.5   Test Results and Comparisons of the Algorithms

Items Algorithm	$\rho(X, X')$	SNR	δ	D_M
ABS	0.991	59.698	1.78	0.012
Adaptive LSB	0.979	59.089	2.85	0.095
LSB	0.955	58.968	2.96	0.125

that the speech quality is closely dependent on the embedding algorithm, of which ABS achieves the best performance and adaptive LSB is better than LSB.

In Table 2.5, δ is less than 3 dB; that is, speech distortion D is controlled within 3 dB ($D < 3$ dB).

The analysis of test results is based on the spectra coming from the original carrier speech, composite speech, and extracted secret speech in G.729. Comparing the spectrum of original carrier speech (Figure 2.4) with that of composite speech (Figure 2.6), analysis shows that our proposed approach keeps the speech's continuity and understandability, and has the features of high capacity and real time. Comparing the spectrum of original secret speech (Figure 2.5) with that of extracted secret speech (Figure 2.7), analysis shows that they are slightly different in signal

FIGURE 2.4

Spectrum of original carrier speech.

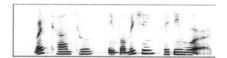

FIGURE 2.5

Spectrum of original secret speech.

FIGURE 2.6

Spectrum of composite speech.

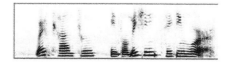

FIGURE 2.7

Spectrum of extracted secret speech.

amplitude and there is a small delay between them. But these defects hardly affect the understanding of the speech. The result of spectrum comparison is satisfied, corresponding to practice listening.

2.5 SUMMARY

It is reasonable to suppose that in the system of secure communication based on the technique of information hiding, no active attack happens and all attacks can be derived from interference of the communication line itself and are generated during transmission. If the communication line of the secure communication system is detected and attacked by the attacker or eavesdropper, it means that the scheme of secure communication based on the technique of information hiding has failed. The simplest way of attacking is to cut off the line of secure communication. It is impossible for an attacker or an eavesdropper to monitor or detect all communication lines, for example PSTN, and from this viewpoint, this hypothesis is reasonable.

This proposed model is well suited for the implementation of practical secure communication systems for the following reasons:

- The speech coding scheme of secret speech and public speech
- The encryption algorithm for secret speech
- The embedding and extraction algorithm

This model is theoretically proved and can be modified flexibly to meet different speech information hiding requirements and different speech coding schemes.
This model can be improved in two ways:

- A more complicated data set is needed to test and improve the performance of this model for the purpose of achieving optimal effects in practical applications.
- The embedding effect of speech quality, hiding capacity, and security can be improved by adopting novel embedding and extraction algorithms.

The ABS Speech Information Hiding Algorithm Based on Filter Similarity

3

Speech information hiding algorithms embed secret speech information in the public speech carrier at the transmitter end for transmission to the receiver through a public channel. Speech information extraction algorithms are used at the receiver end to decode secret speech information that is hidden in the public speech carrier for the purpose of secure communication. Speech information hiding and extraction algorithms are the two core algorithms in the system of secure communication based on the technique of information hiding.

3.1 INTRODUCTION AND MOTIVATION

Speech can be expressed with an autoregressive (AR) model. Each sample X_n is represented as a linear combination of previous L samples plus white noise, e_n, shown in Eq. (3.1) [31,41,122]:

$$X_n = \sum_{i=1}^{L} a_i X_{n-i} + e_n. \tag{3.1}$$

The weights $a_1, a_2, \ldots, a_L$ are called Linear Prediction Coefficients (LPCs).

Samples of input speech are divided into N blocks of samples, called frames. Each frame is typically 10 to 20 ms in length (corresponding to $N = 80 - 160$). Each frame is divided into smaller blocks, of which k samples are equal to the dimension of the Vector Quantization (VQ), called subframes. For each frame, we choose $a_1, a_2, \ldots, a_L$, so that the spectrum of $\{X_1, X_2, \ldots, X_N\}$, generated by using the previous model, closely matches the spectrum of the input speech frames. This is a standard spectral estimation issue and the LPCs $a_1, a_2, \ldots, a_L$ can be calculated by using the Levinson–Durbin algorithm [31,41,122].

3.1.1 BRIEF INTRODUCTION TO THE ABS SCHEME

In an ABS speech coder, for the purpose of generating the same synthetic speech as that in a speech decoder, a speech synthesizer is implemented to combine with a speech analyzer. In order to achieve the minimum error between the synthetic and

Information Hiding in Speech Signals for Secure Communication. DOI: 10.1016/B978-0-12-801328-1.00003-3

original speech, the LPC parameters are adjusted according to the rule of Minimum Square Error (MSE) [123].

The ABS coding scheme procedure is as follows [41,122,123]:

1. Divide speech into frames (about 20 ms length).
2. Analyze the signal of each frame to get the information about LPC parameters and pitch period.
3. Compare the synthetic speech generated by using analysis parameters to synthesis filter with original speech.
4. Adjust the LPC parameters to make the minimum error between synthetic speech and original speech by employing the rule of MSE.
5. Quantize all analysis parameters to prepare for transferring and storing.
6. Regenerate the speech by inputting the excited signal to the LPC filter during the process of decoding.

The basic principles of LPC are introduced next.

3.1.1.1 Basic Principles of LPC

In the speech coding process, direct speech quantization needs a high bit-rate and a large amount of code bits. To ensure high speech quality and to decrease bit-rate, an effective method is to narrow the dynamic range of input signals. LPC uses past samples to predict a new sample value to obtain an error signal. This error signal is generated by subtracting the actual value from the predictive value. Obviously, the dynamic range of error signals is much less than that of the original signals. Hence, the bit-rate of speech quantization could be reduced by using error signals as the input signals to the quantizer [41,122].

Let $s(n)$ $(n=1,2,...,n)$ be the sampled sequence of speech signals. p-order linear prediction uses a weighted sum of past p samples to predict the current value of the signal $s(n)$. Here, the predictor is called the p-order predictor.

Let $\hat{s}(n)$ represent the prediction value of $s(n)$; then [41,122]

$$\hat{s}(n) = \sum_{i=1}^{p} a_i s(n-i),$$ (3.2)

where $a_1, a_2, \cdots, a_p$ are linear prediction coefficients. Equation (3.2) may be called the linear predictor, and p is the order of the predictor.

The transfer function of the p-order predictor is expressed as [41,122]

$$P(z) = \sum_{i=1}^{p} a_i z^{-i}.$$ (3.3)

Let $e(n)$ represent the difference between signal $s(n)$ and its linear prediction value $\hat{s}(n)$; then [41,122]

$$e(n) = s(n) - \hat{s}(n) = s(n) - \sum_{i=1}^{p} a_i s(n-i).$$ (3.4)

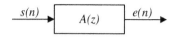

FIGURE 3.1

LPC error filter.

It is easy to see that the error signal $e(n)$ is the output generated by $s(n)$ going through a system $A(z)$ [41,122]:

$$A(z) = 1 - \sum_{i=1}^{p} a_i z^{-i}. \tag{3.5}$$

System $A(z)$ is also called an LPC error filter, which is shown in Figure 3.1 [41,122]. Finding solutions to minimize prediction error $e(n)$ is the task of LPC analysis.

3.1.1.2 ABS Coding Based on LPC

The ABS coding method introduced the synthesizer into the encoder, which is combined with the analyzer. The encoder generates synthesis speech, which is the same as that generated in the decoder. In this method, this synthesis speech is compared with original speech, and each linear prediction coefficient is adjusted to minimize the error signal under a specified criterion. The most widely used criterion is the MSE criterion.

The basic structure of most LPC-ABS codec systems is shown in Figure 3.2 [41,122].

The ABS method first divides speech signals into 20 ms frames. Second, it analyzes each frame to get LPC coefficients (LPCs), pitch period, and other information.

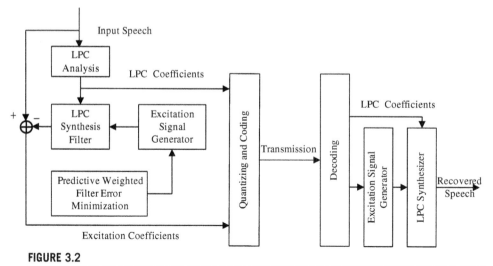

FIGURE 3.2

LPC-ABS codec system.

Then it compares synthesis speech with original speech to adjust parameters. Finally, it quantifies various parameters and transports or stores them. In the decoding end, excitation signals excite the LPC synthesis filter to generate recovered speech.

Many experiments on secure speech communication have been conducted. Chen Liang et al. [118] proposed a speech hiding algorithm based on a speech parameter mode for secure communication. In this approach, the instantaneous pitch is utilized to determine the current embedding positions in public speech (carrier speech) by using Discrete Fourier Transform (DFT) in the frequency domain, and secret information data embedded by modifying relevant frequency coefficients. This approach is very similar to that of watermarking, which has the constraint of hiding capacity and weakness to the attack by low-rate speech coding, such as G.728 and G.729. Anas, Rahman et al. [124] designed a secure telephone prototype called Secure Phone, with practical and reliable encryption, that is easy to operate and connect to PSTN. Secure Phone is a device terminal that is designed to operate reliably, with high speech quality. It can act as both an ordinary telephone and a secure instrument over the dial-up PSTN. Secure Phone operates in full duplex over a single telephone circuit using echo-canceling modem technology [124].

The technique of ABS is widely used in LPC for low-rate speech coding. The principle of the ABS-based speech information hiding algorithm is that it adopts the speech parameters used in ABS coding for secret speech hiding [88].

3.1.2 ANALYSIS OF THE ABS SCHEME

The speech quality is directly correlated to the feature of speech parameters used in ABS coding. The effect of speech parameters on the synthetic speech quality is somewhat different with each coding scheme, although the same type of parameters are used. Experiments on speech quality are conducted by changing different parameters. Results show that for some parameters—called Important Information Bits (IIB), which cannot be used for embedding—speech quality is very sensitive to slight changes. For other parameters—called Tolerant Information Bits (TIB), which can be used to embed secret speech into public speech—it is very sluggish. A slight change to IIBs will lead to great variation of speech quality. Take the linear predictive parameter for example. If there is a very small error in it, the synthetic speech quality will change greatly, with severe degradation. More seriously, small errors that occur in the speech parameter of voiceless and voiced flags could cause the synthetic speech quality to be totally distorted [124].

This analysis specifies that it is very important to select and determine these speech parameters, which have little effect on speech quality when hiding secret messages in public speech by using the speech parameters substitution to realize real-time secure speech communication [50,125].

In fact, speech quality is affected by each speech parameter in ABS coding. So, the simple operation of substituting one parameter with the data bit of secret speech will lead to the degradation of speech quality. For this reason, the operation of embedding secret speech information data bits into TIBs by adopting speech parameters must be taken into consideration during the entire embedding process. First, the TIBs of speech parameters in speech frames must be substituted with secret speech information data bits. Second, an optimal trade-off must be made between speech quality and hiding capacity in the operation of embedding. The best way is to combine the embedding process and the coding process.

Research indicates that speech parameters in traditional LPC-based ABS speech coding schemes are transmitted frame by frame. During this transmission process, the following two issues should be considered for secret speech information embedding [24,88].

3.1.2.1 Frame Similarity

Changes in the features of voiced signal track are very slow during the process of voiced signal generation in speech. The similarity between neighboring frames of the voiced speech segment is very high. This means that some parameters used for generating speech frames are very close. If reasonable degradation of speech quality is accepted, these speech parameters can be transmitted once for two neighboring frames; that is, one frame has parameters and another is vacant. This feature is called frame similarity. From the viewpoint of speech coding, these vacancies could be removed for the purpose of decreasing the coding rate. From the viewpoint of embedding, these vacancies can be substituted with secret speech information data bits for secret speech information hiding [24].

3.1.2.2 Vector Codebooks

The LPC coefficients standing for synthesis filter are quantitative vectors in an LPC-based ABS coding scheme. This method drastically decreases the number of data bits for expressing the LPC coefficients. Traditionally, in speech coding schemes, the design of LPC vectors for all speech samples are stored in a single vector codebook, which is a large table with big storage space and long searching time.

The process of LPC coefficient quantization is a complete coding process to all speech samples. It does not take features of different speech segments into consideration. These features can be adopted by classifying LPC coefficients into different types of groups, and then quantizing these different groups of LPC coefficients into vectors to form multivector codebooks. This process is called LPC coefficients clustering, which results in smaller storage space and shorter searching time for codebooks, and helps to improve efficiency [24,41].

Comparing multicodebooks with single codebooks, experiments indicate that the total storage space of the former is smaller than that of the latter [41,122]. The extra storage space between multi- and single codebooks can be used to store the vectors, whose LPC parameters are substituted with secret speech information data bits in embedding speech frames.

The proposed algorithm of LPCs substitution for designing an ABS speech information hiding approach is based on the previous two conclusions about the feature analysis of an LPC-based ABS speech coding scheme.

The rest of this chapter is organized as follows: Section 3.2 introduces the basic principle of speech information hiding based on the ABS coding scheme, and

Table 3.1   Notations and Abbreviations

Symbol	Definition	Abbreviation	Definition
X_i	i th speech frame	ABS	Analysis-by-synthesis
$m(i)$	Original secret speech frame	AR	Autoregressive
$\hat{m}(i)$	Extracted secret speech	MSE	Minimum Square Error
ξ_i	Normalized correlation coefficient	VQ	Vector Quantization
ξ	Normalized cumulative correlation coefficient	MELP	Mixed Excitation Linear Prediction
Δ_i	Speech variation in energy	CELP	Code Excited Linear Prediction
Δ	Speech cumulative variation in energy	LPCs	Linear Predictive Coefficients
ε_i	Average segment *Itakura* distance	LPC	Linear Predictive Coding
S	Filter similarity coefficient	LSF	Linear Spectrum Frequency
S_0	Filter similarity threshold	TIB	Tolerant Information Bit
R_C	Code rate	IIB	Important Information Bit
R_E	Embed rate	HAS	Human Auditory System
R_P	Public speech code rate	NCCC	Normalized Cumulative Correlation Coefficient
R'_C	Composite speech code rate		

explores two phenomena in the ABS scheme that can be adopted to embed secret speech information. Section 3.3 proposes the concept and calculation of filter similarity, which determines how the LPCs are replaced with secret speech information, and presents the LPCs substitution algorithm and its implementation procedures. Section 3.4 describes the speech information hiding approach based on the ABS scheme for the purpose of secure communication. Section 3.5 explains the setup and configuration of a simulation environment, and performance results. Finally, Section 3.6 summarizes the research findings and comments on further research work.

The notations and abbreviations used in this chapter are summarized in Table 3.1 for convenience [21,22].

3.2 FILTER SIMILARITY

Speeches can be expressed with an autoregressive (AR) model. Each sample X_n is represented as a linear combination of the previous L samples plus a white noise e_n shown in the following equation [31,41,122]:

$$X_n = \sum_{i=1}^{L} a_i X_{n-i} + e_n. \tag{3.6}$$

The weights $a_1, a_2, \ldots, a_L$ are called LPCs.

Samples of input speech are divided into N blocks of samples, called frames. Each frame is typically 10 to 20 ms long (corresponding to $N = 80 - 160$). Each frame is divided into smaller blocks, called subframes, which consist of k samples. The sample number k is equal to the dimension value of Vector Quantization (VQ). For each frame, the LPCs of $a_1, a_2, \ldots, a_L$ are chosen, so that the spectrum of $\{X_1, X_2, \ldots, X_N\}$, generated by using the previous model, matches the spectrum of the input speech frames. This is a standard spectral estimation issue and the LPCs of $a_1, a_2, \ldots, a_L$ can be calculated by using the Levinson–Durbin algorithm [24,31,41,122,123].

A threshold S_0 is set to measure similarity between neighboring LPC synthesis filters. If the similarity between the LPC synthesis filter of the current speech frame, i, and the previous one, $(i-1)$, exceeds the threshold S_0, the transfer of LPC parameters for the i th frame is not necessary, but the parameters for exciting the signal are needed. At the decoding end, the LPC parameters of the $(i-1)$th frame are used for i th frame decoding. This operation is called LPC parameters substitution.

Comparing the quality of speech generated by using the LPC parameters of the $(i-1)$th frame with that using the LPC parameters, results show the two synthetic speeches are very similar, and there is no obvious difference in feeling. The effect caused by the operation of LPC parameters substitution on the speech quality is very small. This means that two speech frames could use the same set of LPCs parameters. Like speech compression, the operation of LPC parameters substitution compresses speech frames on LPCs, and it could decrease the speech coding rate greatly. On the other hand, the empty frame (without LPC parameters) can be used to embed secret speech information into the positions of LPCs parameters. The research in this book focuses on how to adopt filter similarity to hide secret speech information by using empty positions in attentive speech frames [24].

Based on this analysis of frame similarity, the ABS-based information hiding and extraction algorithms are proposed for developing the key technique of secret speech information embedding and extraction, without changing the code rate of the speech coding scheme, but with good speech quality.

For the purpose of determining the LPCs, which will be substituted with secret speech information data bits, the concept of filter similarity is proposed in this book.

Filter similarity is a key quantized indicator to determine the LPC coefficients that can be substituted for embedding secret speech information. The purpose of calculating filter similarity is to measure the capacity of redundant information in speech signals. The calculation results are useful for different types of speech coding standards to select different LPC coefficients for the purpose of embedding secret speech information. The reason for selecting different LPC coefficients is to meet the requirement of different types of speech coding standards according to the speech characteristics [24].

Many kinds of LPC coefficients are used in speech coding. The main speech coefficients of waveform coding are speech sample values, short-term energy, and differential parameters.

The main speech coefficients of the ABS speech coding scheme are [41,122] speech voicing, speech energy, linear prediction (or linear spectral frequency, LSF), excitation, pitch, and various corresponding gain parameters.

Each coefficient reveals a feature of speech coding. These features have a very different impact on the quality of synthetic speech. Even though the same type of coefficients are used in different encoding schemes, the synthetic speech quality is not the same.

For speech coding coefficients, the small change of some coefficients has little impact on the quality of the synthetic speech, but for some other speech coefficients (the IIB) it is not the case. A tiny change on the coefficients would bring a great deviation in synthetic speech. Take the linear predictive coefficients as an example: Even a small error may have a great impact on the quality of synthetic speech, and lead to the severe distortion of the speech. The same is true of the speech voicing mark coefficient. If an error occurs, the synthetic speech is immediately distorted completely [41,122].

Therefore, for the purpose of embedding secret speech information by using the LPC coefficient substitution algorithm, it is necessary to determine the coefficients that have relatively small influence on the speech quality according to different speech coding schemes.

In order to select the most appropriate carrier speech, the mutual relationship between hiding capacity and embedding effect were analyzed, and the results showed that the speech coding schemes GSM (RPE-LTP), G.728 (LD-CELP), and G.729 (CS-ACELP) are more suitable for being selected as public/carrier speech. The better choice for the secret speech coding scheme is MELP and FS1015 (LPC-10e) with a low data rate. The first choice is MELP 2.4 kbps. There are a variety of options for the speech carrier, depending on the applications; for example, the speech coding scheme of G.728 LD-CELP 16 kbps is adopted as public/carrier speech, and the MELP 2.4 kbps scheme is used for secret speech coding [24].

Of course, in the future, with the development of speech compression coding technology, more and more new speech coding schemes with high compression rate and quality will be used in secure communication based on the technology of information hiding.

The filter similarity determination method is implemented based on the LPC ABS speech coding scheme. The basic implementation structure is shown in Figure 3.3 [24,41,122].

In Figure 3.3, each speech segment is composed of many speech frames with a length of 20 ms (160 sample points). Each frame is divided into four subframes with a length of 5 ms (40 sample points).

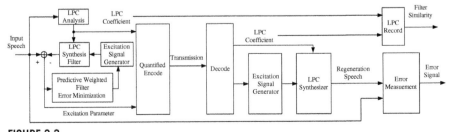

FIGURE 3.3

Filter similarity.

In the realization process, the first step is speech signal segmentation. The procedure is described as follows [24]:

1. Divide speech signals into frames with a length of 20 m. The speech signal is analyzed frame by frame to obtain the LPC coefficients and the pitch information.
2. Generate the synthetic speech in the synthesis filter by using the LPC coefficients obtained through analyzing the speech frame. The synthetic speech is compared with the original speech for the purpose of adjusting the coefficients to achieve the best choice.
3. Quantify the various LPC coefficients, and then the quantified LPC coefficients are transmitted or stored.
4. Produce the synthetic speech in the synthesis filter by inputting the quantized LPC coefficient values as the excitation signal.
5. Compare the synthetic speech with the original speech according to the MES criterion to measure the error level.
6. Record the LPC coefficients if the difference between synthetic speech and original speech is low enough to meet the requirements of secure communication; that is, the error level between two kinds of speech is small enough to be acceptable.
7. Output the results of the filter similarity; that is, the LPC coefficients with minimum errors.

In the LPC-based ABS speech coding scheme, a better way to express the similarity between neighboring frames is with the filter similarity coefficient. The reason is that the wave variation of neighboring frames introduced by the slow changes of the voiced signals track during voiced pronunciation is exhibited mainly on the exciting signal. If two filters have highly similar coefficients, their output waves are very similar with the same exciting signal input. This feature is called filter similarity.

The LPC parameters are the excitation signal (input signal) of the synthesis filter, and the speech frame is the output of the synthesis filter. There are two synthesis filters, f_i and f_{i+1}, with the same LPC parameters as excitation signals that generate two speech frames X_i and X_{i+1}. If X_i and X_{i+1} are similar (frame similarity), the synthesis filters f_i and f_{i+1} should be similar (filter similarity) too. Therefore, filter similarity can be determined by frame similarity in normalized correlation coefficients through adopting perceptually weighted MSE as the fidelity criterion. ξ_i measures the correlation between speech frames X_i and X_{i+1} [104]:

$$\xi_i = \xi(X_i, X_{i+1}) = \frac{\sum_i x_i(i) x_{i+1}(i+1)}{\sqrt{\sum_i x_i^2(i)} \sqrt{\sum_i x_{i+1}^2(i+1)}}, 0 < \xi_i \leq 1. \tag{3.7}$$

Larger ξ_i means higher similarity between speech frames X_i and X_{i+1} in waveform. In this calculation, MSE is used as the criterion to determine the two most similar frames for the purpose of choosing the best LPC parameters to hide secret speech information.

The MSE between two frames is calculated using the following equation [126]:

$$E_i = \sum_{i=1}^{N} [X_i(i) - X_{i+1}(i+1)]^2,$$

(3.8)

where N is the number of the speech frame.

Filter similarity S is calculated according to the following equation [24]:

$$S = \frac{1}{N} \sum_{i=1}^{N} \xi_i.$$

(3.9)

Filter similarity S is the identifier, the trigger point to start the operation of LPCs substitution.

3.3 LPC COEFFICIENT SUBSTITUTION BASED ON FILTER SIMILARITY

In speech communication, like VoIP and PSTN, low-rate speech coding schemes are often used. Hence, in the approach of secure communication over VoIP or PSTN, the filter similarity is used to reduce the number of LPC coefficients transmitted; that is, it is not necessary to transmit LPC coefficients for every speech frame. The omitted LPC coefficients will be replaced by secret speech information bits to realize the secret speech information hiding.

In the speech coding scheme, if some of the LPC coefficients are not transmitted for every speech frame, the equivalent of more redundancy information in speech is deleted. This reduces the data rate of speech coding schemes. It means that a number of data rates can be saved and used for secret speech information transmission. In other words, speech coding scheme data rate R is divided into R_o and R_s. R_o is used for original speech transmission and R_s is used for secret speech information transmission. Hence, the saved data rate defines the hiding capacity.

In order to more clearly explain this phenomenon, a simple example is described as follows. Suppose that a segment speech consists of 16,000 frames in the speech coding scheme with a transmit data rate of 16 kbps. Statistics by calculating the similarity filter show that only 12,000 frames have LPC coefficients, and 1000 frames do not need LPC coefficient transmission. If an average frame occupies three LPC coefficients in a total of 24 bits, this statistic result indicates that 1000 * 24 = 24,000 bits can be used to embed secret speech information, and the secret speech information is transmitted in a average data rate of (16 kbps/1000) * 24 = 384 bps [24].

Using these methods, with the filter similarity threshold S_o being 0.7, the experiments on LPC coefficients substitution have been conducted by testing 36 phrases containing all Chinese consonants and rhyme. Test results show that 36 phrases possess 860 speech frames, in which only 340 speech frames have LPC coefficients. The other 520 frames accounted for 60%, and do not need LPC coefficient transmissions [24].

Take the speech coding scheme of MELP 2.4 kbps as an example. The frame length is 20 ms, and each frame transfers 54 bits, wherein 25 bits are used to transfer LPC coefficients.

If the value of filter similarity S between the synthesis filters of the current frame and one of the frames in front of it is larger than the predetermined value of threshold S_o, it is not necessary to transmit the LPC coefficients of the current frame, in which the LPC coefficients a_0 of the frames in front of it are transmitted to replace a_n.

3.3.1 LPC SUBSTITUTION ALGORITHM

Analysis of the excitation signal in the ABS coding scheme indicates that the similarity between two speech waveform frames is decided by two factors: synthesis filters with greater similarity, and the excitation signal they share. For two synthesis filters with great similarity, if the excitation signal is different, their speech waveforms will be different. Therefore, when the filter similarity of S is greater than S_o, the LPC coefficients of the current frame cannot be transmitted, but still use the signal of the current frame. This guarantees to keep the speech quality of both the current frame and synthetic speech.

The flow chart of the LPC coefficient substitution algorithm based on filter similarity is shown in Figure 3.4 [24].

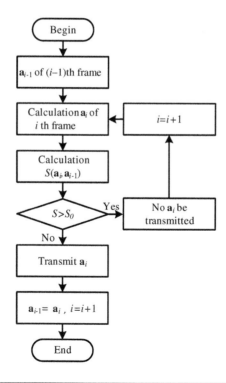

FIGURE 3.4

Flow chart of LPC substitution algorithm.

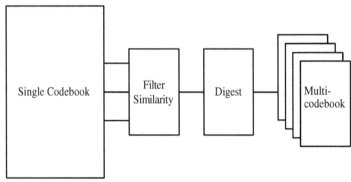

FIGURE 3.5

Multicodebook.

If the filter similarity S between synthesis filters for the i th and $(i{-}1)$th frames exceeds the threshold S_o, the LPC coefficients $\mathbf{a}_i$ for i th frame are replaced by the $\mathbf{a}_{i-1}$ of $(i{-}1)$th frame; that is, substituting $\mathbf{a}_i$ with $\mathbf{a}_{i-1}$. $S(\mathbf{a}_i, \mathbf{a}_{i-1})$ is the similarity of the neighboring i th and $(i{-}1)$th frames, and $\mathbf{a}_i = \mathbf{a}_{i-1}$ means substitution of the LPC parameters $\mathbf{a}_i$ of the i th frame with $\mathbf{a}_{i-1}$ of the $(i{-}1)$th frame.

The threshold S_o is a dynamic value, which depends on speech quality, filter similarity, and other factors. The selection of similarity threshold S_o should guarantee that the LPC substitution algorithm reaches a high hiding capacity (i.e., data rate) and maintains good speech quality.

3.3.2 MULTICODEBOOK

In the LPC substitution algorithm, the multicodebook is established to replace a single codebook, and all LPC parameters are stored in different codebooks separately. The multicodebook in the LPC substitution algorithm is shown in Figure 3.5 [24].

The LPC parameters of the neighboring speech frames are tested to determine the degree of similarity between two speech frames in the function block of filter similarity. If the test result exceeds threshold S_o, as mentioned earlier, only one frame's LPC parameters can be saved, and the LPC parameters of another frame are dropped. Then, a digest is generated as an index of LPC parameters, which are stored in a multicodebook. The result is that just one frame's LPC parameters are kept for two neighboring frames' synthetic speech generation. The memory space in a multicodebook for another frame's LPC parameters is empty, and is prepared for embedding secret speech information.

3.4 SECRET SPEECH INFORMATION HIDING AND EXTRACTION ALGORITHM

Normally, the LPC parameters in the LPC-based ABS coding scheme with a constant code rate R_C are transmitted in each frame. Every speech coding scheme has a constant code rate after compressing coding. If the secret speech is embedded in public speech with an embedding rate R_E, the composite speech is generated with a code rate R'_C (obviously $R'_C = R_C$), which is composed of two code rates: the public speech code rate R_P and the secret speech embedding rate R_E. The relationship among R_C, R_E, and R_P is:

$$R_C = R'_C = R_P + R_E,\tag{3.10}$$

where $R'_C = R_C$ means the public speech is compressed. Comparing the speech quality of public speech with that of composite speech, the former is better than the latter. The degradation of composite speech quality is closely dependent on R_E, and the bigger R_E is, the more degradation to composite speech[1] arises. The ratio of R_E to R_P shows the performance of the proposed approach [22,24,127].

This book adopts the concept of ABS coding to speech information hiding. The ABS approach introduces a speech synthesizer into the speech coding process for information hiding and employs the LPC parameters substitution algorithm to embed secret speech in public speech for composite speech generation.

The speech parameters are calculated and adjusted to minimize errors between composite speech and original speech. A speech synthesizer is used for embedding information data bit streams of secret speech in public speech, which generates the composite speech code stream.

This model for speech information hiding is called the ABS algorithm, and the process of combining the secret speech embedding and composite speech coding in one integrated coding process is called embedded coding. In order to embed secret speech information data bits into a certain speech frame exactly, it is necessary to extract the voiced period in advance. In order to determine the embedding position for hiding secret speech information (i.e., to select the coding parameters for substituting secret speech information), an optimal trade-off among hiding capacity, security, robustness, transparency, and real time must be taken into consideration to achieve the best hiding effect. For this purpose, the embedding depth γ_i is defined as expressing the relationship of all factors about information hiding as an integrated factor [24]:

$$\gamma_i = \xi_i \times \frac{1}{\Delta_i \times \varepsilon_i},\tag{3.11}$$

where $\xi_i\ (0 < \xi_i \leq 1)$ is the correlation coefficient, Δ_i is the variation in energy between frames of public speech and composite speech (Δ_i is an accepted range controlled within 3 dB), and ε_i is average segments of *Itakura* distance (ε_i is an accepted range controlled within 0.036) [120].

The embedding depth γ_i is a dynamic threshold changing with the requirements of information hiding. It is inversely proportional to the product of ε_i and Δ_i. The robustness, hiding capacity, real time, and security of information hiding are improved with the increase of γ_i, although the speech quality and transparency get worse. γ_i may be adjusted to be an optimal value for maintaining good performance of information hiding [24].

---

[1]Secret speech interferes with composite speech and public speech.

3.4.1 SPEECH INFORMATION HIDING ALGORITHM

The algorithm of LPC coefficients substitution for secret speech information hiding based on filter similarity is known as LPC-IH-FS. The schematic diagram of this ABS hiding algorithm is shown in Figure 3.6 [24].

The hiding processes are as follows [24]:

1. To obtain better voice quality, preprocessing is necessary for the original speech carrier. The main purpose is for energy balance and to remove low-frequency interference caused by devices.

2. Divide the original speech into segments according to a certain speech coding method. The segment length depends on the coding algorithm, which is characteristic of the embedding algorithm and its practical requirements. Meanwhile, secret speech is encoded under the prescribed coding algorithm, and is encrypted according to the specified encryption algorithm to ensure data security. Then the encrypted bit stream of secret speech is sent to the buffer.

3. Conduct embedding coding under the specified carrier coding algorithm. In the process of carrier coding, modify the parameter computing method to make the transformed coding parameter meet Eq. (3.11). Compound parameters act as feedback parameters, which are sent to the speech coding algorithm ABS system to participate in the coding process. For all possible $\Phi(s_i, t_i)$, the corresponding compound carrier speech parameters are acquired. The carrier decoder synthesizes the original composite speech by using compound speech parameters. Errors between composite and original speech are calculated in the following equation [24]:

$$E_i = \sum_{i=1}^{M} (t'(t) - t(i))^2. \tag{3.12}$$

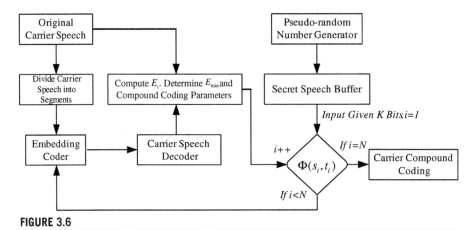

FIGURE 3.6

Schematic diagram of ABS embedding method.

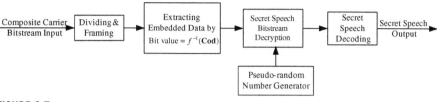

FIGURE 3.7

Schematic diagram of the ABS extraction method.

4. $E_{\min}$ is calculated by [24]

$$E_{\min} = \min\{E_i; i = 1, N\}. \qquad (3.13)$$

Then, secret speech is embedded into public speech by using the embedding method as follows:

$$\Phi(s_{\min}, t_{\min}) \qquad (3.14)$$

The embedding function outputs the composite speech codeword **Cod**.

3.4.2 SPEECH INFORMATION EXTRACTION ALGORITHM

Secret speech information extraction with blind detection-based minimum mean square error (MES) is abbreviated BD-IE-MES. The schematic diagram of this ABS extraction algorithm is shown in Figure 3.7 [24].

The whole process is very simple and fast. It does not need the original carrier speech, which is why it is a "blind" detection process. The extraction process can be illustrated as follows [24]:

1. Divide the received composite bit stream in frames on the basis of the carrier coding algorithm.
2. According to codeword **Cod** and Eq. (3.10), compute the embedded bit value to rebuild the secret speech bit stream, which has not been decrypted yet.
3. Decrypt the extracted bit stream to obtain the decrypted secret speech bit stream. Then the secret speech decoder resynthesizes the secret speech.

3.5 EXPERIMENTAL RESULTS AND ANALYSIS

To achieve better performance of information hiding, the proposed algorithm of LPC substitution based on filter similarity has been tested in a simulation environment connected to PSTN [24,89,90].

3.5.1 SELECTION OF TEST PARAMETERS

The experimental parameters are set as follows.

Frame Length
The speech is divided into frames of 20 ms, consisting of 160 sample points. Each frame is divided into subframes of 5 ms, containing 40 sample points.

Speech Parameters
In the ABS coding scheme, important speech parameters are voiceless speech and voiced speech, speech energy, LPC, Linear Spectrum Frequency (LSF), exciting signal, pitch, and gain.
Tests and measurements show that some of these parameters are changeable and can be used for LPC substitution.

Speech Scheme
The LPC substitution algorithm is suitable for Code Excited Linear Prediction (CELP)-type schemes, such as GSM (Global System for Mobile Telecommunications)-RPE-LTP (Regular Pulse Excitation-Long-Term Prediction), G.728-LD-CELP (Low Delay-Code Excited Linear Prediction), and G.729-CS-ACELP (Conjugate Structure-Algebraic Excited Linear Prediction). Secret speech selects the low-rate or very low-rate coding scheme, such as FS1015 (LPC-10e) and MELP (Mixed Excitation Linear Prediction).
Careful selection of a speech scheme for public speech and secret speech can achieve better performance of information hiding under the constraint of hiding capacity and hiding effect.

Performance Evaluation
Hiding capacity measures the data bits of secret speech, which are embedded into public speech during a certain period. Hiding capacity can express embedding rate in another viewpoint.

Speech variation expresses the difference in quality between public speech and composite speech. The secret speech adopts parameter coding in 2.4 kbps MELP, so its quality cannot simply be measured by the SNR. The normalized cumulative correlation coefficient (NCCC) is used to calculate the degree of secret speech quality degradation of original secret speech $m(i)$ and extracted secret speech $\hat{m}(i)$ according to Eq. (3.8).

Our goal is to achieve an optimal trade-off among speech quality, hiding capacity, and robustness. The criterion for every embedding test is speech quality, which meets the standard of IUT-T P.800 MOS (Mean Opinion Score) and ITU-T G.107 E-Model [122,123], and reaches at least Level 4.

For determining the threshold S_o, an experiment on adopting the LPC coefficient substitution algorithm is conducted by using 36 phrases consisting of 860 speech frames, which contain all Chinese pronunciations. Test results show that 520 out of 860 frames don't need LPC parameters, accounting for about 60%, which means that nearly 60% of memory space in a multicodebook can be saved for embedding secret speech information [24].

Table 3.2   The Test Initial Parameters Setup

Parameters	S_o	γ_i	ξ_i	R_E (bps)
Initial value	0.890	9.259	0.856	360

Take the MELP 2.4 kbps for instance. Each frame has a length of 20 ms, consisting of 54 bits, of which only 25 bits need LPC parameters. That means the other 29 bits can be used for embedding secret speech information.

These experimental results indicate that when the similarity threshold S_o is set to 0.7, the LPC substitution algorithm can reach a hiding capacity (i.e., a data rate greater than 2.4 kbps), and the speech quality may meet the requirements of Level 3.5. If the similarity threshold S_o is set to be greater than 0.8, the hiding capacity may be within the range of 800 to 2.8 kbps, and the speech quality may meet Level 4. Both hiding capacity and speech quality are relatively close to the type of speech coding scheme [24].

In the LPC substitution algorithm, for each speech coding scheme that is chosen as public speech (carrier), the test initial parameters are the same. Based on experimental result statistics, the test initial parameters are provided in Table 3.2.

The test begins after the test initial parameter values are set, then the test software monitors the speech quality variation in energy Δ (which means the change of speech quality) and hiding capacity. During the test, test parameters such as ξ and γ are dynamically adjusted until the best speech quality and hiding capacity are obtained.

3.5.2 EXPERIMENTAL RESULTS

Experiments on hiding capacity and effect by adopting the LPC substitution algorithm to different ABS speech coding schemes are conducted for testing the performance of real-time speech secure communication. In the experiments, public speech is coded in G.721, GSM, G.728, and G.729 schemes respectively, and the secret speech in the MELP 2.4 kbps scheme. The test result of hiding capacity and effect (Table 3.3) shows that G.729 at the rate of 16 kbps achieves a maximum hiding capacity (embedding rate) of 800 bps, which obtains the best performance of speech quality with speech variation of 0.178 dB and a normalized correlation coefficient

Table 3.3   Hiding Capacity and Effect

Scheme	Hiding capacity (bps)	Δ (dB)	ξ	ε	γ
G.721	1600–3600	2.920	0.916	0.029	10.817
GSM	Maximum 2600	2.540	0.945	0.021	17.716
G.728	1600–3200	1.450	0.982	0.018	37.624
G.729	Maximum 800	0.978	0.991	0.015	67.552

Table 3.4 Test Results of Four Traditional Information Hiding Technologies

Item	Adaptive LSB	Phase encoding	Echo coding	Spectrum transform
Hiding capacity (bps)	Maximum 20,000	8–32	10–20, Maximum 346	200
Δ (dB)	3.95	2.24	1.08	2.96

of 0.991. For G.721, the hiding capacity reaches 3600 bps, and the speech quality is worst with the speech variation of 2.920 dB and a normalized correlation coefficient of 0.916 [24].

Optimal trade-off between hiding capacity and speech quality determines that G.728 is the best choice when public speech is used for transferring secret speech.

Experiments on four traditional information hiding technologies have been conducted by using the same public speech and secret speech with ABS. The test results are shown in Table 3.4.

The test results show that:

- Adaptive LSB has the best hiding capacity performance, but its speech variation Δ is the worst. This means Adaptive LSB barters speech quality for hiding capacity.
- Phase encoding has bad performance in hiding capacity and Δ, and is not suitable for real-time communication.
- Echo coding gets a hiding capacity at 346 bps in a good speech quality when the echo delay is controlled in a certain time range.
- Spectrum Transform obtains a hiding capacity at 200 bps, but the speech variation Δ is higher.

Comparing hiding capacity and Δ in Table 3.3 with those in Table 3.4, it is clear that ABS achieves better performance than the other four traditional information hiding technologies.

3.5.3 CALCULATION COMPLEXITY

The execution of an LPC substitution algorithm has three procedures: finding, coding, and hiding. The complexity of computer computations of an LPC substitution algorithm is mainly decided by the selected speech scheme, such as G.721, GSM, G.728, and G.729. The additional computer computations are contributed by the process of finding, which is used for computer filter similarity S and to establish a multicodebook. This step is to find the LPC parameters, which can be substituted by secret speech information. The operation of coding and hiding is a speech synthetic process, which is responsible for speech coding and secret speech information hiding at the same time.

Compared with the original speech scheme, experiments show that the LPC substitution algorithm increases the calculation complexity by less than 5%.

3.5.4 **SPEECH QUALITY**

The precise measurement of speech quality is a complicated process. Traditionally, the test method of speech quality is subjective. In practical applications, the wave and spectrum of speech signals are two elements that can be used to measure speech quality.

In our proposed approach, the test results from public speech and composite speech's wave and spectrum and the secret speech signal's waves and spectra are shown in Figures 3.8, 3.9, 3.10, and 3.11 respectively.

Comparison of the waves and spectra of speech and the original public speech with the composite speech shows that the speech features are well maintained, and the speech quality for communication is guaranteed. The differences between public speech and composite speech are:

- **Slight change:** Due to secret speech information data bits being embedded in public speech, there are some changes between the frame data of composite speech and public speech.
- **Small delay:** Small delay is introduced by adopting the LPC substitution algorithm to perform the operations of calculating the positions and embedding data bits.

Comparing original secret speech with extracted secret speech in wave and spectrum show that there are slight changes and small delays in speech frame data, with good speech quality guaranteed because the lossless extraction method is used.

3.6 **SUMMARY**

According to the characteristics of ABS speech coding, different types of speech coding schemes, such as G.711, G.721, G.728, G.729, and GSM, are chosen as the public speech carrier, and the MELP with a data rate of 2.4 kbps is selected as the secret speech. The algorithm of LPC coefficient substitution for secret speech information hiding based on filter similarity (LPC-IH-FS) and information extraction algorithm with MES-BD are proposed.

The main topics covered in this chapter are the following.
Filter similarity
This chapter puts forward the concept of filter similarity, and gives out its definition and quantitative calculation method. Filter similarity is a good way to present the comparability in different frames of the speech signal.
For two great similarity filters, their output waveform is very similar to each other when the same excitation signal is used to drive two filters individually.
The dullness pronunciations with a slow variation in sound characteristics and the changes in the waveform between two neighboring frames are decided by the excitation signal. Normally, the synthesis filters of two frames have great similarity.

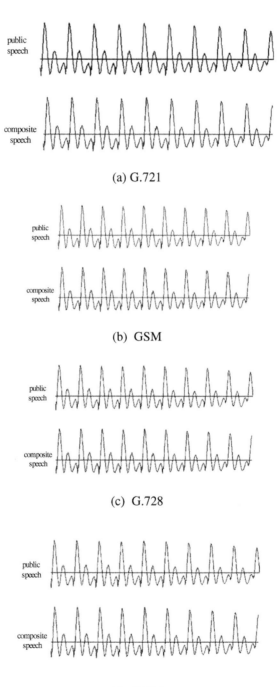

FIGURE 3.8

Experimental results of public speech and composite speech in waves.

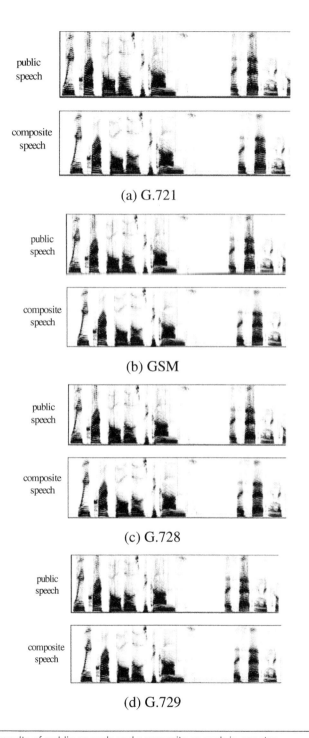

FIGURE 3.9

Experimental results of public speech and composite speech in spectra.

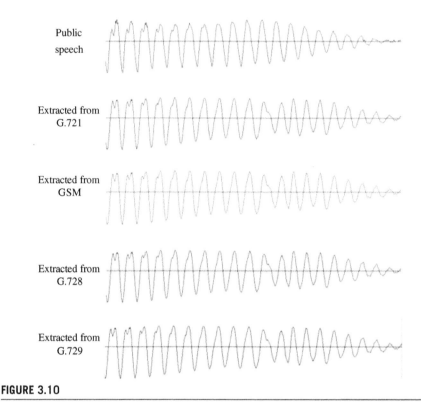

Public speech

Extracted from G.721

Extracted from GSM

Extracted from G.728

Extracted from G.729

FIGURE 3.10

Experimental results of secret speech in waves.

The approach of LPC coefficients substitution based on filter similarity

When the similarity between two LPC synthesis filters of current and one neighboring speech frame is greater than a predetermined threshold, the excitation signal parameters are transmitted to substitute LPC coefficients of the current speech frame.

The LPC synthesis filter of the current speech frame is replaced by the ones of the previous speech frame during the process of decoding. This operation generates synthetic speech, which is less different from the ones generated by using the LPC synthesis filter of the current speech frame. This replacement has less effect on synthetic speech quality, but can greatly reduce the code rate. Filter similarity and the LPC substitution algorithm are used to implement the approach of speech hiding and extraction, which embeds secret speech information in a public speech carrier without perceptible speech quality degradation. The embedded secret speech information cannot be detected during its transmission over the network. In case the data rate of communication network is constant, this approach can provide enough hiding capacity to ensure the continuity of communication speech. Therefore, the proposed approach achieves the real time, robustness, and security in secure communication.

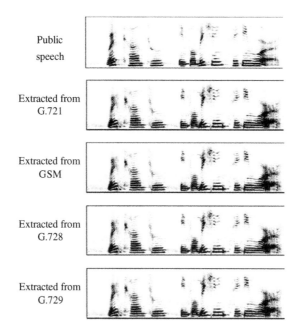

Public
speech

Extracted from
G.721

Extracted from
GSM

Extracted from
G.728

Extracted from
G.729

FIGURE 3.11

Experimental results of secret speech in spectra.

LPC-IH-FS and BD-IE-MES algorithms

The algorithm of LPC coefficient substitution for secret speech information hiding based on filter similarity (LPC-IH-FS) and information extraction with BD-IE-MES are proposed.

In the proposed algorithm, the speech synthesizer is introduced into the decoder, in which the synthesizer is combined with the analyzer to generate a synthetic speech that is completely consistent with the synthetic speech generated in the encoder. Comparing the synthetic speech with the original speech, the minimum error is achieved by calculating and adjusting all the parameters in the proposed algorithm according to a predetermined error criterion, for example MES.

In the proposed speech coding based information hiding scheme, the speech synthesizer is introduced to embed the secret speech bit stream in the public speech bit stream and generate the composite speech. The embed process and speech coding fusion mean that the operations of secret speech embedding and public speech coding are carried out together. In the multicode or multiframe embedded case, the composite speech stream and the original speech respectively are decoded to generate the synthetic speech.

The embedding method with different parameters is used to hide secret speech; different composite speeches are generated. All the generated composite speeches are compared with original speech and the error between composite and original speech is calculated to obtain the parameters of embedding method with minimum error; that is, the best performance embedding method is determined. In the case of single frame embedding, the proposed algorithm makes full use of the synthetic function of carrier speech coding to realize the secret information embedding.

This work presents an LPC substitution algorithm based on the ABS speech coding scheme to design an ABS speech information hiding approach for the purpose of real-time speech secure communication. The LPC substitution algorithm uses the LPCs as the carrier to carry secret speech information data bits to embed secret speech in public speech.

The performance of the LPC substitution algorithm is evaluated by adopting different ABS speech coding schemes in secure communication over VoIP and PSTN. Experimental results show that the proposed algorithm meets the requirements of secure communication in real time, security, and embedding rate.

The analysis of test results shows that in the LPC-based ABS coding scheme, the similarity between two neighboring frames is not only decided by the filter similarity, but also depends on the exciting signal. Hence, when the exciting signals for two high-similarity filters are the same, the two neighboring frames are very similar. If two different exciting signals are input to two high-similarity filters, the two neighboring frames are different. The conclusion is that there is a small effect on the speech quality of the output frame by using the exciting signals of the current frame under the condition of $S > S_0$, and there are no LPC parameters being transmitted. The degradation of speech quality is in an acceptable range.

For wide practical application, the LPC substitution algorithm needs more and more complicated data to test for and to improve the performance in the future.

The G.721-Based Speech Information Hiding Approach

G.721 is an International Telecommunications Union (ITU-T) standard for speech compression and decompression that is used in digital transmission systems, and in particular, it is widely used in stream media communication and telecommunication [128].

This chapter applies the idea of ABS-based speech information hiding to the AD-PCM speech coding scheme to realize the G.721-based secure communication. ABS-based speech information hiding and extraction algorithms are applied to the G.721 based on the approach proposed in Chapter 3. In this chapter, a brief introduction to the G.721 speech coding scheme is given first and then detailed embedding and extraction operations are described.

4.1 INTRODUCTION TO THE G.721 CODING STANDARD

G.721 is an ITU-T standard codec with channel. G.721 uses adaptive differential pulse code modulation (ADPCM) to generate a digital signal with a lower bit rate of a 32 kbit/s than standard PCM. G.721 was first introduced in 1984. In 1990 this standard was folded into G.726 along with G.723 [49,128].

To better understand the G.721 speech coding standard, it is necessary to learn the basis of differential pulse code modulation first.

4.1.1 DIFFERENTIAL PULSE CODE MODULATION

Differential pulse code modulation (DPCM) is based on the notion of quantizing the prediction-error signal. In many signal sources of interest, such as speech, samples do not change a great deal from one to the next; in other words, the samples are correlated with their neighbors. If the current sample can be predicted from previous samples, it is possible to form the prediction-error signal, with significantly lower variance and dynamic range. By quantizing the prediction error, a higher signal-to-noise ratio (SNR) can be achieved for a given resolution. The technique is shown in Figure 4.1 [128].

In Figure 4.1, the prediction error $e(n)$ is obtained by subtracting the input $x(n)$ from the prediction $x_p(n)$, which is quantized. The indices at the output of the quantizer's encoder represent the DPCM bit-stream. The indices are entered into the

Information Hiding in Speech Signals for Secure Communication. DOI: 10.1016/B978-0-12-801328-1.00004-5

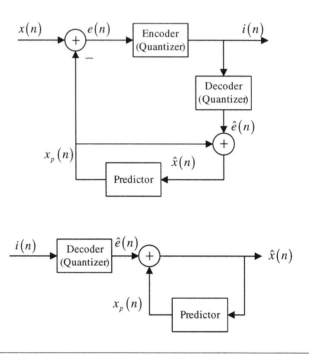

FIGURE 4.1

DPCM encoder (top) and decoder (bottom).

quantizer's decoder to obtain the quantized prediction error, which is combined with the prediction $x_p(n)$ to generate the quantized input. The DPCM decoder works in a similar fashion to obtain the quantized samples from the indices [128].

Note that prediction is based on the quantized signal samples, which is indeed suboptimal since higher performance can be achieved using the original, unquantized signal samples. The approach is utilized mainly because the decoder has no access to the original input, and synchronization must be maintained between the encoder and the decoder.

4.1.2 ADAPTIVE SCHEMES

In scalar quantization, adaptation is necessary for optimal performance when dealing with nonstationary signals like speech, where properties of the signal change rapidly with time. These schemes are often referred to as APCM [128].

4.1.2.1 Forward Gain-Adaptive Quantizer

Forward adaptation can accurately control the gain level of the input sequence to be quantized, but side information must be transmitted to the decoder. The general structure of a forward gain-adaptive quantizer is shown in Figure 4.2 [128].

A finite number N of input samples (frame) are used for gain computation, where $N \geq 1$ is known as the frame length. The estimated gain is quantized and used to scale

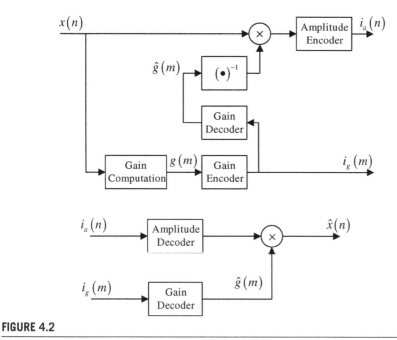

FIGURE 4.2

Encoder (top) and decoder (bottom) of the forward gain-adaptive quantizer.

the input signal frame; that is, $x(n) / \hat{g}(m)$ is calculated for all samples pertaining to a particular time. Note that a different index m is used for the gain sequence, with m being the index of the frame. The scaled input is quantized with the index $i_a(n)$ and $i_g(m)$ transmitted to the decoder. These two indices represent the encoded bit-stream. Thus, for each frame, N indices $i_a(n)$ and one index $i_g(m)$ are transmitted. If transmission errors occur at a given moment, distortions take place in one frame or a group of frames; however, subsequent frames will be unaltered. With sufficiently low error rates, the problem is not serious [128].

Many choices are applicable for gain computation. Some popular schemes are [128]

$$g(m) = k_1 \max_n \left\{ |x(n)| \right\} + k_2 \tag{4.1}$$

$$g(m) = k_1 \sum_n x^2(n) + k_2. \tag{4.2}$$

With the range of n pertaining to the frame associated with index m, and (k_1, k_2) being positive constants, the gain is used to normalize the amplitude of the samples inside the frame, so that high-amplitude frames and low-amplitude frames are quantized optimally with a fixed quantizer. To avoid numerical problems with low-amplitude frames, k_2 is incorporated so that divisions by zero are avoided.

4.1.2.2 Backward Gain-Adaptive Quantizer

In a backward gain-adaptive quantizer, gain is estimated on the basis of the quantizer's output. The general structure is shown in Figure 4.3 [128].

Such schemes have the distinct advantage that the gain need not be explicitly retained or transmitted since it can be derived from the output sequence of the quantizer. A major disadvantage of backward gain adaptation is that a transmission error not only causes the current sample to be incorrectly decoded but also affects the memory of the gain estimator, leading to forward error propagation.

Similar to the case of the forward gain-adaptive quantizer, gain is estimated so as to normalize the input samples. In this way, the use of a fixed amplitude quantizer is adequate to process signals with wide dynamic range. One simple implementation consists of setting the gain $g(n)$ proportional to the recursive estimate of variance for the normalized-quantized samples, where the variance is estimated recursively with [128]

$$\sigma^2(n) = \alpha\sigma^2(n-1) + (1-\alpha)y^2(n), \tag{4.3}$$

where $\alpha < 1$ is a positive constant. This constant determines the update rate of the variance estimation. For faster adaptation, set α close to zero. The gain is computed with

$$g(n) = k_1\sigma^2(n) + k_2, \tag{4.4}$$

where k_1 and k_2 are positive constants. The constant k_1 fixes the amount of gain per unit variance. k_2 is incorporated to avoid division by zero. Hence, the minimum gain is equal to k_2.

In general, it is very difficult to analytically determine the impact of various parameters (α, k_1, k_2) on the performance of the quantizer. In practice, these parameters are determined experimentally depending on the signal source.

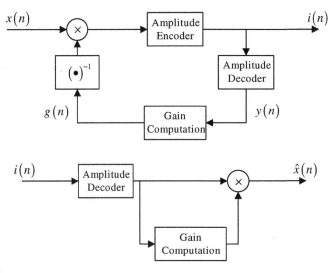

FIGURE 4.3

Encoder (top) and decoder (bottom) of the backward gain-adaptive quantizer.

4.1.2.3 *Adaptive Differential Pulse Code Modulation*

The DPCM system described earlier has a fixed predictor and a fixed quantizer; much can be gained by adapting the system to track the time-varying behavior of the input. Adaptation can be performed on the quantizer, on the predictor, or on both of them. The resulting system is called adaptive differential PCM (ADPCM) [50,85]. The encoder and the decoder of an ADPCM system with forward adaptation is shown in Figure 4.4 [128].

As for the forward APCM scheme, side information is transmitted, including gain and predictor information. In the encoder, a certain number of samples (frame) are collected and used to calculate the predictor's parameters. For the case of the linear predictor, a set of LPCs is determined through LP analysis. The predictor is quantized with the index $i_p(m)$ transmitted.

As in DPCM, prediction error is calculated by subtracting $x(n)$ from $x_p(n)$. A frame of the resultant prediction-error samples is used in gain computation, with the resultant value being quantized and transmitted. The gain is used to normalize

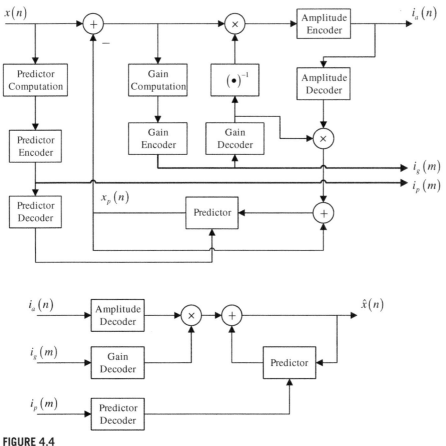

FIGURE 4.4

Encoder (top) and decoder (bottom) of a forward-adaptive ADPCM quantizer.

the prediction-error samples (samples of normalized prediction error, gain, and the predictor's parameters), which are used in the encoder to compute the quantized input $\hat{x}(n)$, and the prediction $x_p(n)$ is derived from the quantized input. This is done because, on the decoder side, it is only possible to access the quantized quantities. In this way, synchronization is maintained between the encoder and the decoder since both are handling the same variables.

One shortcoming of the forward-adaptation scheme is the delay introduced due to the necessity of collecting a given number of samples before processing can start. The amount of delay is proportional to the length of the frame. This delay can be critical in certain applications since echo and annoying artifacts can be generated.

Backward adaptation is often preferred in those applications where delay is critical. Figure 4.5 [128] shows an alternative ADPCM scheme with backward adaptation [50,85].

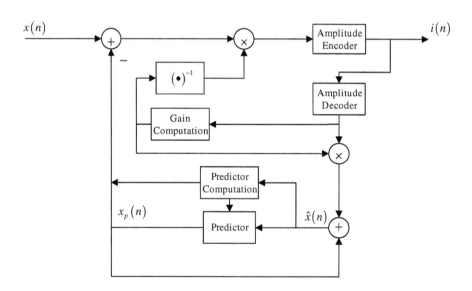

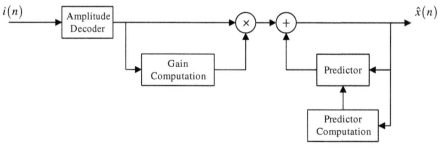

FIGURE 4.5

Encoder (top) and decoder (bottom) of a backward-adaptive ADPCM quantizer.

Note that the gain and predictor are derived from the quantized-normalized prediction-error samples; hence, there is no need to transmit any additional parameters except the index of the quantized-normalized samples.

Similar to DPCM, the input is subtracted from the prediction to obtain the prediction error, which is normalized, quantized, and transmitted. The quantized-normalized prediction error is used for gain computation. The derived gain is used in denormalization of the quantized samples; these prediction-error samples are added with the predictions to produce the quantized input samples. The predictor is determined from the quantized input $\hat{x}(n)$. Techniques for linear predictor calculation are given in Chapter 1. Using recursive relations for the gain calculation of Eqs. (4.3) and (4.4) and linear prediction analysis, the amount of delay is minimized since a sample can be encoded and decoded with little delay. This advantage is mainly due to the fact that the system does not need to collect samples of a whole frame before processing. However, be aware that backward schemes are far more sensitive to transmission error since they affect not only the present sample but also future samples due to the recursive nature of the technique [128].

The standard of G.721 ADPCM is the technique of waveform coding, which compresses data adopting the high correlation between samples and adaptive step-size. It can accomplish the mutual conversion between the rate of 64 kbps PCM in A or μ law and 32 kbps. The input of the coder and decoder of G.721 ADPCM is G.711 PCM code, which is expressed by 8 bits with a sample rate of 8 kHz; hence, the data rate of G.711 is 64 kbps. The G.721 ADPCM code is generated by the adaptive quantizer. This code expresses the differential signal in 4 bits with the same sample rate of 8 kHz. The data rate of G.721 ADPCM is 32 kbps. The data compression ratio of G.721 and G.711 is 2 to 1. Thus, the G.721 scheme can provide high compression ratio but ordinary sound quality [128,129].

4.2 THE APPROACH TO HIDE SECRET SPEECH IN G.721

Based on the analysis of G.721 speech coding standard, the approach of embedding secret speech into the G.721 coding scheme and extracting secret speech from the G.721 coding scheme is proposed. In this section, the embedding process is illustrated first, followed by the corresponding extraction process.

4.2.1 EMBEDDING ALGORITHM

In this approach, the secret speech is coded by 2.4 kbps MELP, while the public (carrier) speech is coded by 32 kbps G.721-ADPCM. The basic principle of the G.721-based information hiding and extraction algorithm is to import the linear PCM coding sampled points of original speech into the G.721-ADPCM coder one by one, and embed the data bits of secret speech information during the process of coding (i.e., the processes to embed and encode are executed synchronously).

Before the secret speech is embedded, the preprocessing of original speech consists of two procedures: (1) band-pass filtering, to remove the low-frequency noise introduced by MIC or any other devices, and (2) energy balance, because too much energy of input speech leads to overflow during adaptive quantization of embedding coder. The process of normalizing input speech energy is gained based on the benchmark, which is the maximum value of short-term speech energy, and may prevent overflow from happening. Preprocessing is very important to speech quality and embedding effect. Quantization overflow will produce strong noises, which may degrade the speech quality seriously, although in the meantime it may also lower the hiding capacity and weaken the robustness [92].

Public speech is divided into frames after preprocessing, and the length of each frame is determined by the embed rate. The length is calculated by the equation:

$$Frame\_len = 8000 / embed\_rate, \tag{4.5}$$

where *embed_rate* is the rate for embedding secret speech information into public speech. If *embed_rate* equals 2 kbps, the *Frame_len* is four codewords. Hence, the composite speech can be expressed as $S[i], i = 1, \dots, 4$, which means only one bit of secret speech is embedded in each frame. Embedding operations are shown in Figure 4.6 [92].

The input public speech of PCM in *A* or μ law is converted into uniform PCM. The differential signal $d(k)$ is obtained by calculating the error value, which is the

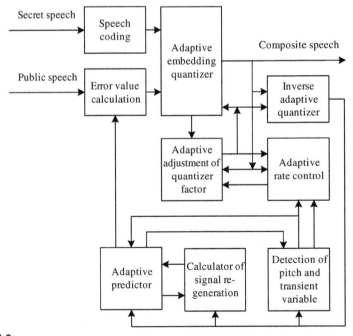

FIGURE 4.6

The encoder of the embedding process.

difference between input speech $x(k)$ of uniform PCM and the predictive signal $x_e(k)$ (estimation value), $d(k) = x(k) - x_e(k)$. The adaptive embedding quantizer has two functions:

- It is used as a quantizer to express the differential signal $d(k)$ in 4 bits (but 0000 is eliminated).
- It is used as a synthesizer to embed secret speech into public speech.

The inverse adaptive quantizer generates the differential signal $d(k)$ according to 4-bit code that is used in the adaptive quantizer. The input signal $x(k)$ is reconstructed by the differential signal $d(k)$ plus the predictive signal $x_e(k)$. An adaptive predictor generates the predictive signal of input speech using the reconstructed signals $x(k)$ and $d(k)$. This process is an adaptive feedback loop, which automatically makes the error between public speech and composite speech reach the minimum under the rule of Mean Square Error (MSE). This means that secret speech is embedded into public speech with minimum error, which introduces the least degradation to the quality of speech [92].

The flow chart of embedding secret speech into public speech is shown in Figure 4.7 [92].

For convenience, the notations used are explained as follows [92]:

N_embed is the number of secret speech bits.
Italic CW is a codeword of speech, bold **CW** is a vector that consists of one or more codewords.
$\mathbf{CW}_i$ is the i th codeword vector, $i = 1, 2, 3, 4$.
Italic BV is the value of each bit in codeword CW, called bit value, and BV_i is the i th bit value, $i = 1, 2, 3, 4$.
Bold **BV** is a vector that consists of one or more bit values.
W is the sum of a 4-bit value with weight control.
W_i is the weight of the i th bit, $i = 1, 2, 3, 4$.
C is the hiding capacity.
Φ_i is the embedding method, and $i = 1, 2, \ldots, 80$.
S is the sample point of public speech.
S_P is the sample point of composite speech.
f is the embedding function, which describes the relationship of **CW** and embedded **BV**.
f^{-1} is the extraction function.

If $C = 1$ bit, the $f(\mathbf{CW},\mathbf{BV}) = 0$ is obtained, then $\mathbf{BV} = f^{-1}(\mathbf{CW})$.

The embedding operation is controlled by the error between public speech and composite speech under the constraint of the MSE rule. Detailed operation steps are as follows [92]:

1. Divide the public speech into frames.
   a. The public speech is divided into frames, which means that the number of frames depends on the length of public speech. And the length of a frame, which corresponds to the hiding capacity, is determined by Eq. (4.5).

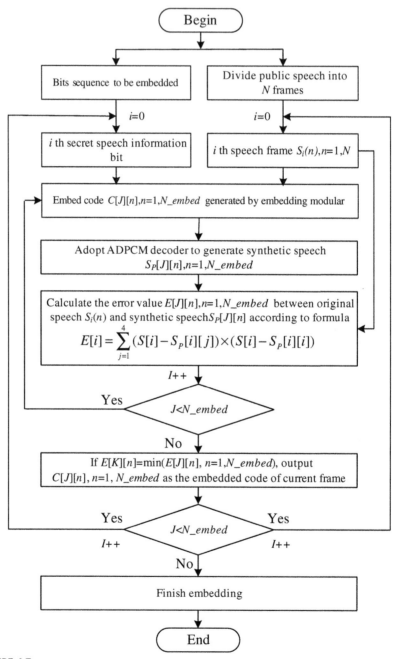

FIGURE 4.7

Flow chart of embedding and coding.

2. Split the adaptive quantity table.
   a. The adaptive quantity table *TABLE* [8] stores the quantity value in the form of an 8-bit codeword *CW*. For better embedding and convenient extracting, *TABLE* [8] is divided into two quantity tables of *TABLE0* [4] and *TABLE1* [4]; that is, the 8-bit codeword is split into two parts, each with 4 bits.

   Definition 1: The codewords generated from *TABLE0* [4] have the weight of even number and the codewords generated from *TABLE1* [4] have the weight of odd number.

   Definition 2: The relationship between embedding function *f* and the summation $W = \sum W_i \oplus BV_i$ satisfies the equation

$$f(\mathbf{CW}_i, \mathbf{BV}; i = 1,2,3,4) = \sum_{i=1}^{4} W_i \oplus BV = 0. \tag{4.6}$$

3. Select the embed method according to the embedding group, determine the number of embedded bits, and record its change

   Experiments for *FRAME_len* = 4 show that if the LSB of each frame is used for transformation, the total state number can be calculated as [121,130]

$$2^1 \times c_4^1 + 2^2 \times c_4^2 + 2^3 \times c_4^3 + 2^4 \times c_4^4 = 80.$$

   Therefore, the total number of embedding method Φ_i is 80; the *i* th transform is noted as Φ_i (*i* = 1,...,80).

   Each Φ_i sends one frame of public speech sample points and one bit of secret speech to the embedding encoder for embedding and coding, which is performed according to Eq. (4.6). The operation parameters in this process and embedding codewords $\mathbf{CW}[i][j]$ ($i = 1,2,\ldots,80, j = 1,\ldots,4$) are recorded by Φ_i for embedding and coding. Each group $\mathbf{CW}[i][j]$ is coded in the ADPCM encoder, which generates synthetic speech sample points $S_p[i][j]$ of 80 groups.

   It is well known that the process of embedding will somewhat degrade the quality of speech. The difference between synthetic speech and public speech is calculated by the minimum error value:

$$E[i] = \sum_{j=1}^{4} (S[i] - S_p[i][j]) \times (S[i] - S_p[i][i]). \tag{4.7}$$

   Every $E[i]$ corresponds to one Φ_i, and the minimum error value is determined by

$$E[k] = \min\{E[i]; i = 1,80\}. \tag{4.8}$$

   The embedding method Φ_K corresponding to the minimum error $E[k]$ is determined as the selection for ADPCM coding and embedding. It generates codewords $\mathbf{CW}[k][j]$ ($j = 1,\ldots,4$).

   The parameters of the ADPCM encoder and decoder are updated by using the correlative variables of embedding method Φ_K for the next frame encoding.

4.2.2 EXTRACTION ALGORITHM

In the G.721 decoder, some parts of the components are the same as the feedback loop of the encoder. Additionally, it includes the components of PCM conversion and synchronous coding adjustment. PCM conversion is used for transforming the compressed PCM in A or μ law to PCM. The synchronous coding adjustment prevents the distortion of accumulation in the process of synchronous serial coding [92].

When the secret speech is embedded into ADPCM encoded public speech, the *Frame_len* = 4 and the embedding encoder outputs the codewords in the same frame under the conditions of Eq. (4.8). The decoding and extraction process are executed synchronously to retrieve secret speech.

Extracting secret speech is a simpler process (Figure 4.8) than embedding, and no side information about secret speech or public speech is required [92].

The embedding position and variables are updated to the decoder for extraction in time.

According to the embedding process, *Frame_len* = 4 is taken as an example.

1. Divide the composite code stream data into frames.

    a. When composite code stream data are received at the receiver end, they are divided into frames according to the allocated length *Frame_len*.

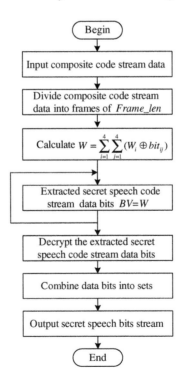

FIGURE 4.8

Flow chart of secret bits extraction and decoding.

2. Calculate W.

    a. The value W is calculated by using the equation $W = \sum_{i=1}^{4} \sum_{j=1}^{4} \left(W_i \oplus bit_{ij} \right)$; it represents the embedded bit value of secret speech (i.e., $BV = W$). This means that once the value of W is obtained, the embedded bit value BV is extracted.

    b. The process of calculating W corresponds to the extraction of secret speech, which is executed for calculating W one frame by one frame until complete secret speech information is totally extracted.

3. Decode secret speech.

    a. The operation of secret speech decoding recovers the data stream of secret speech to form the original format, and then decodes the extracted bit to retrieve the secret speech. That's the end of the extraction process.

4.3 EXPERIMENTAL RESULTS AND ANALYSIS

The experiment of embedding secret speech of 2.4 kbps MELP into ADPCM speech is conducted in the environment of two connected computers equipped with a speech process card, with which the speech algorithm and speech information hiding are implemented. Tests mainly focus on hiding capacity and speech quality [92].

4.3.1 HIDING CAPACITY

Hiding capacity is a quantity, which indicates how many secret speech bits are embedded into public speech. It is restrained by hiding effect and calculation complexity. The computational complexity is closely associated with the hiding effect. Obviously, the more complicated the calculation is, the higher the hiding capacity is.

Experimental results (Table 4.1) show that the hiding capacity for embedding secret speech into ADPCM varies from 1.6 kbps to 2.0 kbps with no need for the side information of original public speech and secret speech. Meanwhile, the decoding calculation is not complicated, and the steganography, security, and real time are guaranteed

If the convenience of extraction is not considered, the secret speech embedding rate can reach 8.13 kbps with good hiding effect [92].

Table 4.1   Experimental Results

Hiding capacity	1.6 kbps	2.0 kbps	8.13 kbps
Variation	1.05 dB	1.36 dB	3.45 dB
Calculation complexity	Low	Middle	High

4.3.2 SPEECH QUALITY

The degradation of speech quality can be measured by the variation between original speech and synthetic speech in Power Spectrum Density (PSD). For each embedding method Φ_i, the error between the sample points of original speech and synthetic speech is calculated to find the best way to embed the secret speech. The error for each Φ_i (i.e., each operation of embedding) is variable. The average change of errors is measured in decibels. The smaller the variation, the better the quality of synthetic speech. Here synthetic speech refers to composite speech.

Table 4.1 shows that the variation is associated with hiding capacity. The high capacity introduces bigger errors and bigger speech variation, which means more degradation of the speech quality.

In experiments, the speech waveform and spectrum can directly indicate the performance of speech information hiding. Secret speech is a short sentence, "Welcome to speech information hiding world," and public speech is a free talk between two extremely voluble young women. Test results give the waveform and spectrum of public, secret, and composite speech (Figures 4.9–4.16) at 1600 bps embedding rate [92].

Because the process of extraction is lossless, comparison of original, composite, and secret speech in waveforms and spectra indicates that the basic speech features are well preserved, and although there is a slight change to the spectrum and a little delay introduced, no distinct speech quality degradation occurred. Analysis of the

FIGURE 4.9

Original secret speech waveform.

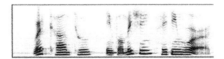

FIGURE 4.10

Original secret speech spectrum.

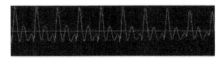

FIGURE 4.11

Original public speech waveform.

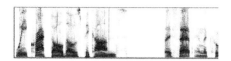

FIGURE 4.12

Original public speech spectrum.

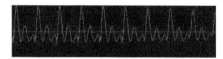

FIGURE 4.13

Composite speech waveform.

FIGURE 4.14

Composite speech spectrum.

FIGURE 4.15

Extracted secret speech waveform.

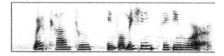

FIGURE 4.16

Extracted secret speech spectrum.

speech spectra and waveforms shows that this approach of speech information hiding has a high hiding capacity of 1600 bps with an excellent speech quality and a good property of speakers' recognition.

Experiments on four classical information hiding technologies have been conducted by using the same public speech and security speech with G.721. Test results are displayed in Table 4.2 [92].

Table 4.2   Test Results for Four Classical Information Hiding Technologies

Item	Adaptive LSB	Phase coding	Echo coding	Spectrum transform
Hiding capacity (bps)	Maximum 20,000	8–32	10–20, Maximum 346	200
Variation (dB)	3.95	2.24	1.08	2.96

Test result shows that:

- Adaptive LSB has the largest hiding capacity, but its speech variation is also the biggest. This means Adaptive LSB barters speech quality for hiding capacity.
- Phase coding has a bad performance in hiding capacity and speech variation, and it is not suitable for real-time communication.
- Echo coding has a maximum hiding capacity of 346 bps with a good speech quality while the echo delay is controlled in a certain time range.
- Spectrum transform has a hiding capacity of 200 bps, but the speech variation is bigger.

Comparing Table 4.1 with Table 4.2, it is clear that the proposed ABS approach has better performance than the four traditional information hiding technologies.

4.4 SUMMARY

In this chapter, an approach of secure communication based on the technology of information hiding is realized on concrete speech coding according to the model of information hiding. The G.721-based secure communication approach is presented. In this approach, the speech coded in the G.728 scheme is used as public/carrier speech and a 2.4 kbps MELP speech is selected as secret speech. The proposed approach succeeds in conducting a secure communication at 32.0 kbps with a secret speech embed rate of 1.6 kbps based on the algorithm of LPC-IH-FS. The extraction of secret speech information is completed by using the algorithm of BD-IE-MES. In this approach, public speech as a carrier plays a role in two aspects: one, it provides a common communication channel for public information; two, it sets up a subliminal channel for secret speech (steganography information).

The proposed approach has been adopted by Speech Information Hiding Telephony (SIHT), which has been tested in a PSTN environment, and a satisfactory result has been achieved. The proposed approach can be modified flexibly to satisfy different speech information hiding requirements, as well as different speech coding schemes.

The drawback of this proposed approach is that the algorithm is a little complicated, and more calculation is needed.

The G.728-Based Speech Information Hiding Approach

5

G.728 is an ITU-T standard for speech coding operating at 16 kbit/s. It is officially described as coding of speech at 16 kbit/s using low-delay code excited linear prediction (LD-CELP). The G.728 algorithm has become one of the required algorithms in VoIP. G.728 is used mainly in many fields of VoIP, satellite communication, and voice storage, among others [48].

This chapter applies the idea of analysis-by-synthesis (ABS)-based speech information hiding to the LD-CELP speech coding scheme to realize the G.728-based secure communication. ABS-based speech information hiding and extraction algorithms are applied to the G.728 based on the approach that was proposed in Chapter 3. In this chapter, a brief introduction to the G.728 speech coding scheme is given first and then detailed embedding and extraction operations are presented.

5.1 CODE EXCITED LINEAR PREDICTION

CELP has been the most widely used speech coding algorithm over the past 10 years. But CELP is commonly used as a generic term for a class of algorithms and not for a particular codec [131].

The essence of CELP techniques, which is an ABS approach to codebook search, is retained in LD-CELP, which uses backward adaptation of predictors and gain to achieve an algorithmic delay of 0.625 ms.

The CELP algorithm is based on four main ideas [131]:

- Using the source-filter model of speech production through linear prediction (LP)
- Using an adaptive and a fixed codebook as the input (excitation) of the LP model
- Performing a search in a closed loop in a "perceptually weighted domain"
- Applying vector quantization (VQ).

5.1.1 THE CELP SPEECH PRODUCTION MODEL

The CELP coder relies on the long-term and short-term linear prediction models. Figure 5.1 shows the block diagram of the speech production model, where an excitation sequence is extracted from the codebook through an index [48,131].

Information Hiding in Speech Signals for Secure Communication. DOI: 10.1016/B978-0-12-801328-1.00005-7

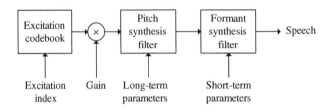

FIGURE 5.1

The CELP model of speech production.

The extracted excitation is scaled to the appropriate level and filtered by the cascade connection of pitch synthesis filter and formant synthesis filter to yield the synthetic speech. The pitch synthesis filter creates periodicity in the signal associated with the fundamental pitch frequency, and the formant synthesis filter generates the spectral envelope [48,131].

The codebook can be fixed or adaptive and can contain deterministic pulses or random noise. For simplicity, the speech production model that the CELP coder relies on consists simply of a white noise source exciting the synthesis filters.

The CELP coding method is based on the ABS algorithm, introduced in Chapter 1. Figure 5.2 shows the block diagram of an encoder with the closed-loop approach [48,131].

Conceptually, this closed-loop optimization procedure's goal is to choose the best parameters so as to match the synthetic speech with the original speech as much as possible.

5.1.2 CODING PRINCIPLES

Speech signal is represented by a combination of parameters: gain, filter coefficients, voicing strengths, and so on [41]. Theoretically, all parameters of the speech coder can be optimized jointly so as to yield the best result. However, it is very complex due to the computation involved. In practice, only a subset of parameters is selected for closed-loop optimization, while the rest are determined through an

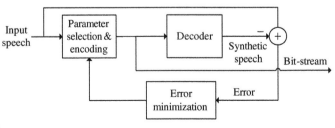

FIGURE 5.2

Block diagram showing an encoder based on the ABS principle.

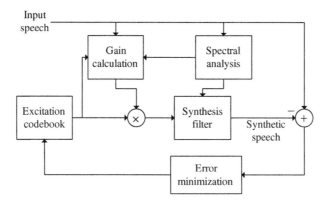

FIGURE 5.3

Block diagram showing the key components of a CELP encoder.

open-loop approach. The CELP coder is based on the ABS principle, where the excitation sequences contained in a codebook are selected according to a closed-loop method. Other parameters, such as the filter coefficients, are determined in an open-loop fashion. Figure 5.3 illustrates a simplified block diagram of the CELP encoder [48].

A commonly used error criterion, such as the sum of squared error, can be applied to select the final excitation sequence; hence, waveform matching in the time domain is performed, leading to a partial preservation of phase information [132].

Since the model requires frequent updating of the parameters to yield a good match to the original signal, the analysis procedure of the system is carried out in blocks. That is, the input speech is partitioned into suitable blocks of samples to determine the time-varying of filter parameters. The excitation signal reaches optimal when the time-varying is fixed. The excitation sequence and other parameters are then encoded to form a compressed bit-stream. The length of the analysis block or frame affects the bit-rate of the coding scheme [41].

5.1.3 ENCODER OPERATION

A block diagram of a generic CELP encoder is shown in Figure 5.4. This encoder is highly simplistic and serves only as an illustration. The encoder works as follows [41,48,132]:

1. The input speech signal is segmented into frames and subframes. The scheme of four subframes in one frame is a popular choice. Length of the frame is usually around 20 to 30 ms, while for the subframes it is in the range of 5 to 7.5 ms.
2. Short-term LP analysis is performed on each frame to yield the linear predictive coefficients (LPCs). Afterward, long-term LP analysis is applied to each subframe.

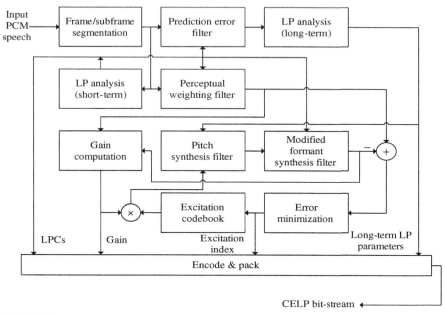

FIGURE 5.4

Block diagram of a generic CELP encoder.

Input to short-term LP analysis is normally the original speech, or preemphasized speech; and input to long-term LP analysis is often the (short-term) prediction error. Coefficients of the perceptual weighting filter (discussed later), pitch synthesis filter, and modified formant synthesis filter are given after this step.

3. The excitation sequence can now be determined. The length of each excitation code vector is equal to that of the subframe; thus, an excitation codebook search is performed once every subframe. The search procedure begins with the generation of an ensemble of filtered excitation sequences with the corresponding gains; mean-squared error is computed for each sequence, and the code vector and gain associated with the lowest error are selected.

4. The index of excitation codebook, gain, long-term LP parameters, and LPCs are encoded, packed, and transmitted as the CELP bit-stream.

5.1.4 PERCEPTUAL WEIGHTING

The perceptual weighting filter is based on the masking effect of the human audio system. In speech, the noise that is carried in a higher energy frequency band (around formant) is less easily perceived than the one carried in a lower energy frequency band. Thus, this conclusion can be used to measure the difference between original speech and synthetic speech, allowing a relatively large difference in a higher energy frequency band.

One efficient way to implement the weighting filter is to use the system function [41,132]:

$$W(z) = \frac{A(z)}{A(z/\lambda)} = \frac{1 - \sum_{i=1}^{r} a_i z^{-i}}{1 - \sum_{i=1}^{p} a_i \lambda^i z^{-i}}. \tag{5.1}$$

The character of the perceptual weighting filter is determined by prediction coefficients $\{a_i\}$ and weighting factor $\lambda \in |0,1|$. The constant λ determines the degree to which the error is deemphasized in any frequency region. The filter amplifies the error signal spectrum in nonformant regions of the speech spectrum, while attenuating the error signal spectrum in formant regions. Hence, an error signal whose spectral energy is concentrated in formant regions of the input spectrum is considered better than the one whose spectral energy is not located under formants.

From Eq. (5.1), $W(z) \to 1$ can be derived as $\lambda \to 1$. Hence, no modification of the error spectrum is performed. On the other hand, if $\lambda \to 0$, $W(z) \to A(z)$, which is the formant analysis filter. The constant λ introduces a broadening effect (bandwidth expansion) to the error weighting filter. The most suitable value of λ is selected subjectively by listening tests, and for 8 kHz sampling λ is usually between 0.8 and 0.9 [132].

5.1.5 VECTOR QUANTIZATION

Vector quantization (VQ) is a kind of signal compression method. CELP coding uses the VQ method to compress data, such as an excitation signal, LPCs, and codebook gain. VQ concerns the mapping in a multidimensional space from a (possibly continuous-amplitude) source ensemble to a discrete ensemble. The mapping function proceeds according to some distortion criterion or metric employed to measure the performance of VQ [41,132].

The VQ technique usually includes the establishment of a codebook and the search of codewords. The establishment of a codebook, also known as the codebook training, generally requires that the training set is 50 times that of a codebook. The LBG iteration algorithm is a commonly used method to train the codebook. Splitting training is utilized to avoid divergent calculation that is caused by improper selection of the initial codebook. Additionally, large codebook training needs a large amount of computation and storage capacity.

5.2 INTRODUCTION TO THE G.728 CODING STANDARD

Figures 5.5 and 5.6 display the encoder and decoder block diagram of the G.728 scheme respectively [48].

The work principle of encoding is transforming the input signal to a uniform PCM signal at first. The input signal can be A law or μ law PCM signal at the rate of 64 kbit/s. Then a five-dimensional speech vector is established in an encoding

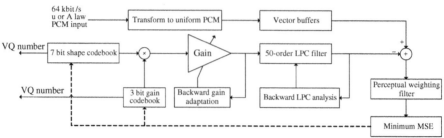

FIGURE 5.5

Block diagram for G.728 LD-CELP encoder.

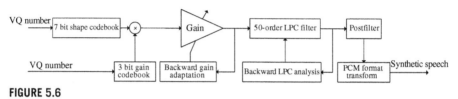

FIGURE 5.6

Block diagram for G.728 LD-CELP decoder.

process by using five consecutive speech sample points. An excitation codebook contains 1024 five-dimensional code vectors, and the number of code vectors (10 bits) is sent to a decoder. Every four adjacent input vectors (20 samples) constitute an adaptive cycle, or a frame. LPCs are updated by every frame. Because of the adoption of backward adaptive prediction in LD-CELP, the current excitation gain is obtained by the previous quantized gain and the output of the synthesis filter is obtained by LPC analysis of the speech information. The information sent to the decoder is merely the address number of the excitation vectors. Hence, there exists only a five-sample delay, which means 0.625 ms delay for the sampling rate of 8 kHz. Taking handling and transmission delay into consideration, the total delay of a one-way codec does not exceed 2 ms [41,48].

The decoding operation is carried out in a one-by-one vector. The corresponding excitation vector is found in the excitation codebook according to the received code vector number. After gain adjustment, the excitation signal can be obtained, which will be transmitted into the 50-order LPC filter to build synthetic speech. This synthetic speech goes through the postfilter to reinforce the subjective sensation quality of speech.

CELP and LD-CELP are ABS-LPC-based speech coding, and share the following common characteristics [41,48]:

- Estimating and analyzing the LPC synthesis filter parameters in an open loop
- Adding local decoder and perceptual weighting filter in the encoder end
- Searching the codebook in a closed loop to get the optimal stimulus parameters under the optimal rule so that the mean of perceptual weighting error is minimized

In order to achieve low-delay and high-quality speech coding, LD-CELP has the following different characteristics:

- Employing backward adaptive prediction. By analyzing the past output of the quantified value of the local decoder, the LPCs can be obtained and used for the current frame operating linear prediction.
- Estimating the envelope of the speech signal spectrum and its detailed structure accurately by using a 50-order short-term predictor in the absence of a long-term (pitch) predictor.
- Adopting a hybrid analysis window. The window function consists of recursive and nonrecursive parts. It is easy to calculate the autocorrelation function by the recursive method.

5.3 THE CELP-BASED SCHEME OF SPEECH INFORMATION HIDING AND EXTRACTION

The approach to hide secret speech in a CELP-type coding scheme is realized by adopting the ABS speech information hiding and extraction algorithm [133].

5.3.1 EMBEDDING SCHEME

The realization of ABS speech information embedding in a CELP-type coding scheme integrates the process of information embedding into a CELP coder and synthesizer, as shown in Figure 5.7 [133].

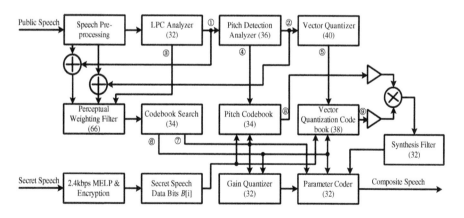

① Short-term Speech Information   ② Long-term Speech Information   ③ 10 LSP Parameters Per 240 Samples

④ Optimum Pitch Index and Optimum Pitch Gain Per 60 Samples   ⑤ Optimum Code Book Index and Optimum Code Book Gain Per 60 Samples

⑥ Short-term Speech Residual   ⑦ Long-term Speech Residual   ⑧ Pitch Analysis   ⑨ Vector Quantization Codebook Searching

FIGURE 5.7

CELP-based speech information embedding.

Public speech is typical human speech, which is a uniformly quantized PCM signal at a rate of 16 bits with a sample rate of 8 K. The preprocessing of a PCM signal involves eliminating the noise and balancing the energy. The LPC analysis and perceptual weighting filtering produce short-term and long-term speech information. It uses an organized, nonoverlapping, deterministic algebraic codebook containing a predetermined number of vectors, uniformly distributed over a multidimensional sphere to generate a remaining speech residual. The short-term and long-term speech information and remaining speech residual are combinable to form a quality reproduction of the digital speech input [118,133,134].

In known CELP coding techniques, each set of excitation samples in the codebook must be used to excite the LPC filter and the excitation results must be compared using an error criterion. Normally, the error criterion used determines the sum of the squared differences between the original and the synthetic speech samples resulting from the excitation information for each speech frame. These calculations involve the convolution of each excitation frame stored in the codebook with the perceptual weighting impulse response. Calculations are performed by using vector and matrix operations of the excitation frame and the perceptual weighting impulse response [39,133,135,136].

The codebook is constructed by uniformly distributing a number of vectors over a multidimensional sphere. This is accomplished by constructing ternary valued vectors (where each component has the value −1, 0, or +1), having 80% of their components with value zero, and fixed nonzero positions. The fixed position of the nonzero elements is uniquely identifiable with this coder in comparison with other schemes [39,133,135,136].

Codebook ID is the lookup address for speech frames. A pitch codebook ID consists of 34 bits, which divides into two subframes; the second subframe searches individually. A vector quantization codebook ID is 38 bits, which is composed of two subframes, each 19 bits.

Not all parameters can be used for secret speech information hiding. Experiments show that a slight modification to an LSP (Linear Spectrum Pair) may result in great harm to the speech quality. This is because a modification to the LSP coefficient changes all sampled points that pass through the linear predictor. The degree of change is dependent on the input sampled points, and it cannot be estimated before. Especially during the period of transition, when speech intensity changes from weaker to stronger or from stronger to weaker, it will lead to a greater change of sample points. In fact, the period of transition is a very important part in the speech, carrying more information [118,134,135]. It may bring great change to some sampled points, which creates a great difference in speech quality between the original speech and the synthetic speech. This similarity to the parameters of adaptive code vector gain and fixed code vector gain and the fluctuation of their values greatly affects synthetic speech quality, and it introduces more noise into the synthetic speech. Hence, the parameters mentioned earlier cannot be used for secret information hiding.

Experimental results show that the pitch error introduced by a slight modification to the ID of the pitch codebook is proportional to pitch frequency, and has a small effect on the synthetic speech. So, the ID of the pitch codebook is one parameter that can be used for information hiding. The results also suggest that public speech must be carefully selected to achieve optimal information hiding results when adopting the ID of a pitch codebook as the embedding parameter in the proposed ABS information hiding approach. For example, the public speech selects the man's speech, whose pitch frequency is lower.

Another parameter selected as the embedding parameter in the proposed ABS information hiding approach is the ID of the vector quantization codebook. Generally, it is better than the ID of the pitch codebook in the quality of synthetic speech. Like the previous analysis and experiment, the ID of the pitch codebook and ID of the vector quantization codebook are determined as the embedding parameters used for secret information hiding in the proposed approach.

Based on a CELP-type speech coding scheme, in the process of embedding, one frame of input data is composed of two parts: one is the sampled points of public speech, and the other is a group of 8 bits of secret information bits $B[i]$, $i = 0,...,7$. The vector quantization codebook is generated by its algebraic construction. Its selection depends on whether it meets the selection requirements. The gain of the pitch codebook and the vector quantization codebook are vector quantities in conjugating construction [133].

The procedures of embedding are as follows [133]:

1. Divide each frame into subframes, which consists of certain sampled points.
2. Analyze every sampled point in each frame using LP, and convert them to LSF (Linear Spectrum Frequency) parameters.
3. Quantize and code the vector into a different number of bits according to the selected public speech coding scheme.
4. Construct perceptual weighted filter $W(z)$ and synthesis filter $H(z)$ individually using quantitative and nonquantitative LP parameters.
5. Select the pitch codebook and vector quantization codebook under the control of embedding data sets $B[i]$ and predetermined constraint function F.
6. Multiply the two selected exciting signals by their own gains individually, and add the two results together to get a new exciting signal as the input to the synthesis filter $H(z)$, which generates the local reconstruct signal $\tilde{S}(n)$.
7. Calculate the perceptive weighted MSE (Minimum Square Error) between $S(n)$ and $\tilde{S}(n)$ to determine the minimum MSE, and make the exciting signal optimization.
8. Output the current frame's codeword CW of the optimal exciting signal by the embedding encoder. This codeword consists of the signal's ID in the codebook, gain, LP parameters, and other parameters [133].

The operation of one frame embedding is finished. By repeating this procedure, more secret speech frames can be embedded.

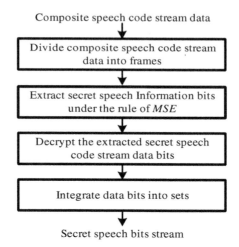

Composite speech code stream data

Divide composite speech code stream data into frames

Extract secret speech Information bits under the rule of *MSE*

Decrypt the extracted secret speech code stream data bits

Integrate data bits into sets

Secret speech bits stream

FIGURE 5.8

CELP-based speech information extraction.

The function F constrains sets, which makes the one-to-one relationship between CW and $B[i]$:

$$B[i] \underset{F^{-1}}{\overset{F}{\rightleftharpoons}} CW,$$

where CW is an output codeword generated by embedding the coder used for the current frame. $B[i]$ is embedded data bits.

5.3.2 EXTRACTION SCHEME

At the end of the receiver, the ABS extraction algorithm is adopted to a CELP-type coding scheme for extracting the secret speech information. The CELP-based speech information extraction scheme is shown in Figure 5.8 [133].

The code stream data of composite speech are divided into frames. The embedded secret speech data bits that can be extracted depend on the relationship between F of $B[i]$ and CW considering the rule of MSE, without side information of original speech. After the encryption of extracted secret speech code stream data, the data bits are integrated into sets, which is the secret speech bits stream that can be transmitted from the transmitter to receiver.

The extracted secret speech bits stream coded in 2.4 kbps MELP form the synthetic secret speech.

5.4 APPROACH TO HIDE SECRET SPEECH IN G.728

This section is focused on the security of communication channels, and proposes an approach of real-time speech secure communication based on the technique of information hiding. This approach embeds 2.4 kbps MELP-coded secret speech into

CELP G.728-type coded speech by adopting the ABS-based algorithm of LPC-IH-FS and BD-IE-MES [118,134]. This approach embeds, transmits, and extracts secret speech information for the purpose of secure communication.

5.4.1 EMBEDDING ALGORITHM

The flow chart of the embedding algorithm for G.728-based carrier speech is shown in Figure 5.9 [118,133,134].

Carrier speech $x(n)$ is a uniform quantized PCM signal, forming a vector (or subframe) every five consecutive samples. It is well known that the output of the G.728 coding scheme is the number of optimal vectors under MSE rule conditions. The length of an optimal codebook is 10 bits and the codebook contains 1024 independent vectors, stored in a vector table. To reduce the complexity of a codebook search, the codebook is split into two vector tables [26,48].

The first table is a shape codebook, which contains 128 independent code vectors and is 7 bits long. The second one is a gain codebook, containing eight zero-symmetrical scalarized values. The first bit represents the flag and other two bits express the altitude.

From a large number of experimental results, it can be concluded that the fluctuation of codebook gain will generate a big noise, which leads to serious speech quality degradation. In terms of shape codebook vectors, selecting part of a codebook for embedding coding according to a certain rule may not have a great impact on the quality of synthetic speech. Hence, the shape codebook vectors are chosen as embedding parameters. As different embedding capacities under various conditions and security requirements are required, how to determine the function f between embedding parameters and embedded bits is a very important issue.

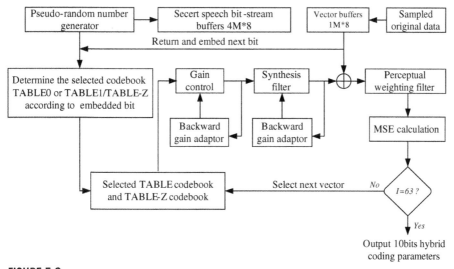

FIGURE 5.9

Flow chart of embedding algorithm.

In accordance with the provisions of coding, every five consecutive samples make up the input of a subframe and every four subframes comprise frame data. For each input subframe, an embedding encoder selects a code vector in turn based on an embedded bit [41,122]:

- If the embedded bit is 0, choose the code vector from **TABLE0** and **TABLE-Z**.
- If the embedded bit is 1, choose the code vector from **TABLE1**.

Here, **TABLE-Z** is the original gain codebook. **TABLE0** and **TABLE-Z** are split from the shape codebook, and they each contain 64 independent code vectors. The selected code vector satisfies the predestinate constraint function $f(\mathbf{X},\mathbf{Bit}) = 0$, where $\mathbf{X}$ refers to the 10-bit number of the vector, and **Bit** expresses the functional relationship between embedding parameters and embedded bits. f^{-1} is the extraction function, which satisfies

$$\mathbf{Bit} = f^{-1}(\mathbf{X}). \tag{5.2}$$

The selected code vector is imported into an excitation synthesis filter after being marked by the gain, forming a local decoded signal. Compute the frequency weighted mean square error between the local signal and original signal, and then choose the code vector that retains the minimum error as the optimal code vector. Finally, send the 10 bits of the optimal code vector, which are the last results of the embedding coding. LPCs are updated in each frame. Extract and update the excitation gain one-by-one vector by utilizing the previous quantized gain of the excitation signal.

The frequencies weighted mean square error is computed as [122,133]:

$$MSE = \left\| x(n) - \tilde{x}_{ij} \right\| = \sigma^2(n) \left\| \hat{x}(n) - g_i H y_j \right\|, \tag{5.3}$$

where H stands for the cascade system function of the synthesis filter and perceptual weighting filter. g_i is the i th gain value of **TABLE-Z**, y_j is the j th code vector of **TABLE0** or **TABLE1**, and $\hat{x}(n) = x(n) / \sigma(n)$.

5.4.2 EXTRACTION ALGORITHM

A block diagram of the extraction algorithm of G.728-based speech hiding is shown in Figure 5.10 [48,133].

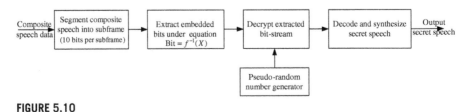

FIGURE 5.10

Block diagram of extraction.

Due to the output codewords, X of the embedding encoder satisfies constraint function $f(\mathbf{X},\mathbf{Bit}) = 0$, and it is very convenient to extract secret data without any original carrier speech information. Such an extraction method is also called *blind* detection.

First, the segment receives composite speech into a subframe (10 bits per subframe) to form a codeword X, then secret information bits are extracted using Eq. (5.2). Finally, the bit-stream is extracted and decoded to rebuild the secret speech.

5.5 EXPERIMENTAL RESULTS AND ANALYSIS

Two computers equipped with 16-bit sound cards and speech processing cards are used by connecting via PSTN (Public Switched Telephone Network) to set up a test experimental environment. The sound card is used for speech input, output, and monitoring. The speech processing card implements speech coding and embedding.

Public speech is recorded in PCM, and coded in a CELP-type speech coding scheme. It is a segment of normal speech used as a carrier for secret speech. Secret speech is recorded in PCM, and coded in 2.4 kbps MELP. Public speech is repeatable and used for embedding full secret speech. It is divided into small sections of 20 ms, consisting of four 5 ms frames.

The experiment adopts G.728 LD-CELP with the data rate of 16 kbps as the speech coding scheme of public speech. Secret speech is the phrase "Welcome to the information hiding world" spoken by a middle-aged man. The data bits of secret speech coded in 2.4 kbps MELP are embedded in G.728 one-by-one corresponding with the parameters of G.728. These parameters have been modified, processed, and substituted with secret speech data bits; that is, embedding secret speech information data bits in public speech. The extraction of secret speech data bits strictly corresponds to embedding [134].

Experiments show that an average data rate of 3200 bps can be embedded into public speech in real time with good performance in hiding capacity, speech quality, and secret message steganography (Table 5.1) [134].

The difference between normal speech and synthetic composite speech is measured by speech variation in decibels.

Table 5.1 Performance Test Results of ABS Algorithm

Compress coding scheme	G.728
Hiding capacity (bps)	1600–3200
Speech variation (dB)	1.45

FIGURE 5.11

Original public speech spectrum.

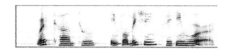

FIGURE 5.12

Original secret speech spectrum.

FIGURE 5.13

Composite speech spectrum.

FIGURE 5.14

Extracted secret speech spectrum.

By comparing spectra of original public speech (Figure 5.11) with composite speech (Figure 5.12), analysis shows that our proposed approach keeps the speech's continuity and understandability, and has a high capacity and real time. Analyzing the spectra of original secret speech (Figure 5.13) and extracted secret speech (Figure 5.14) shows that they are slightly different in signal amplitude and have small delay, but these defects hardly affect the understanding of the speech. The result of the spectrum comparison is satisfactory, corresponding with actual hearing [134].

5.6 SUMMARY

In this chapter, an approach of secure communication based on G.728 is presented. In this approach, speech coded in the G.728 scheme is used as public/carrier speech and a 2.4 kbps MELP speech is selected as secret speech. The proposed approach realizes secure communication at 16.0 kbps with secret speech being embedded at the rate of 1.6 to 3.2 kbps by using the LPC-IH-FS algorithm. The secret speech information is extracted using the BD-IE-MES algorithm. The public speech coded in the G.728 scheme is used to build two common communication channels: one is an open channel used for public speech transmission, and the other is a subliminal channel used for secret speech (steganography information) communication.

The proposed approach is adopted by SIHT, and has been satisfactorily tested in a PSTN environment.

The G.729-Based Speech Information Hiding Approach

6

G.729 is an audio data compression algorithm for voice that compresses digital voice in packets of 10 ms duration. Recommendation ITU-T G.729 contains the description of an algorithm for the coding of speech signals using conjugate-structure algebraic-code-excited linear prediction (CS-ACELP) [51].

Because of its low bandwidth requirements, G.729 is used mostly in VoIP applications where bandwidth must be conserved, such as conference calls. Dual-tone multifrequency signaling (DTMF), fax transmissions, and high-quality audio cannot be transported reliably with this codec [51].

This chapter applies the idea of ABS-based speech information hiding to the CS-ACELP speech coding scheme to realize G.729-based secure communication. ABS-based speech information hiding and extraction algorithms are applied to G.729 based on the approach that was proposed in Chapter 3. In this chapter, a brief introduction to the G.729 speech coding scheme is given first and then the detailed embedding and extraction operations are introduced.

6.1 INTRODUCTION TO THE G.729 CODING STANDARD

G.729 is officially described as coding of speech at 8 kbit/s using CS-ACELP, but there are extensions that provide rates of 6.4 kbit/s and 11.8 kbit/s for worse and better speech quality respectively. G.729 has been extended with various features, commonly designated as G.729a and G.729b [51].

6.1.1 ALGEBRAIC CODEBOOK STRUCTURE

As studied in Chapter 5, a CELP coder contains a fixed excitation codebook holding the code vectors that serve as input to the synthesis filter. The codebook is searched during encoding to locate the best code vector for a particular speech subframe. The G.729 coder has an algebraic structure for the fixed codebook (referred to from now on as the algebraic codebook) that is based on the interleaved single pulse permutation design. In this scheme each code vector contains four nonzero pulses. Each pulse can have the amplitude of either 1 or −1 and can assume the positions [50,51].

Each excitation code vector has 40 samples, which is the length of a subframe. The excitation code vector $v(n)$ is constructed by summing the four pulses according to [41,51]

$$v[n] = \sum_{i=0}^{3} p_i[n] = \sum_{i=0}^{3} s_i \delta[n - m_i]; \quad n = 0, \dots, 39, \tag{6.1}$$

where $s_i = \pm 1$ is the sign of the pulse and m_i is the position. Thus, each pulse requires 1 bit per sign. For p_0, p_1, and p_2, three bits are needed for position; for p_3, four bits are required. Hence, a total of 17 bits are needed to index the whole codebook.

Therefore, the index to the codebook is composed of two parts, the sign of four bits, and the position of 13 bits [41,51].

The codebook index is completed by using a truncation function,

$$f(x) = x, \text{ if } x > 0$$
$$f(x) = 0, \text{ otherwise.}$$

The sign index is given by

$$\sin dex = f(s_0) + 2f(s_1) + 4f(s_2) + 8f(s_3) \tag{6.2}$$

and is represented by four bits.

The position index is obtained by

$$pindex = \frac{m_0}{5} + 8\left\lfloor \frac{m_1}{5} \right\rfloor + 64\left\lfloor \frac{m_2}{5} \right\rfloor + 512\left\lfloor 2\frac{m_3}{5} + ms_3 \right\rfloor \tag{6.3}$$

and is represented by 13 bits, where

$$ms_3 = \begin{cases} 0; & if \quad m_3 = 3,8,\dots,38 \\ 1; & if \quad m_3 = 4,9,\dots,39 \end{cases}. \tag{6.4}$$

Note that Eqs. (6.2) and (6.3) are the actual indices transmitted as part of the G.729 bit-stream [51]. The advantage of this method is that no physical storage is required for the fixed codebook. However, the amount of bits allocated is relatively higher when compared with other CELP standards.

6.1.2 ADAPTIVE CODEBOOK

The G.729 coder has some unique methodologies applied to the adaptive codebook. In general, the G.729 is more complex with finer structure to boost the quality of synthetic speech. Due to advances in DSP technology, extra complexity can be tolerated.

6.1.2.1 Perceptual Weighting Filter

G.729 utilizes a more sophisticated perceptual weighting filter (Figure 6.1) than previous generations of CELP standards [51].

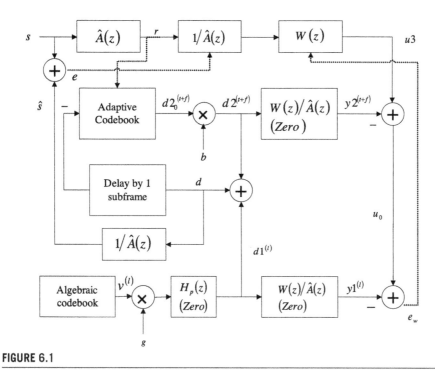

FIGURE 6.1

Signal involved in the G.729 encoder.

The system function is [41,51]

$$W(z) = \frac{A(z/\gamma_1)}{A(z/\gamma_2)} = \frac{1 + \sum_{i=1}^{10} a_i \gamma_1^i z^{-i}}{1 + \sum_{i=1}^{10} a_i \gamma_2^i z^{-i}}. \tag{6.5}$$

The LPC a_i is the unquantized coefficient and is updated once every subframe. For each subframe, the LPCs are obtained through interpolation in the LSF domain.

The parameters γ_1 and γ_2 determine the frequency response of the filter and are made adaptive depending on the spectral characteristics. The adaptation is based on a spectrum flatness measure obtained through the first two RCs, as well as the closeness of the LSFs. There are two possible values for γ_1—0.94 or 0.98—and γ_2 pertains to the range [0.4, 0.7]. The rules for the determination of γ_1 and γ_2 are based on subjective tests, and improvement in final speech quality has been reported.

6.1.2.2 Open-Loop Pitch Period Estimation

Open-loop pitch period estimation is the first step toward adaptive codebook search. The procedure is based on the weighted speech signal and is done once per frame; that is, every 10 ms. Figure 6.2 [51] summarizes the major steps involved.

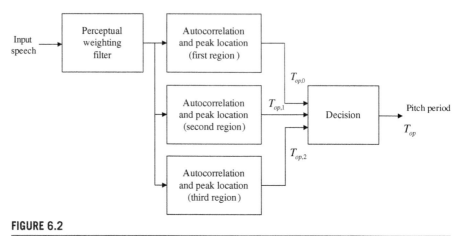

FIGURE 6.2

The open-loop pitch period estimation procedure for the G.729 encoder.

The perceptually weighted speech, denoted by $u(n)$, is used in the following autocorrelation expression [41,51]:

$$R[l] = \sum_{n=0}^{79} u[n]u[n-l].$$ (6.6)

Note that 80 samples are involved, corresponding to two subframes. Three peaks are separately found in three ranges of lag. These results are denoted by [41,51]

$$R_0 = R[T_{op,0}] \mid R[l] \le R[T_{op,0}], \; 20 \le l, T_{op,0} \le 39$$ (6.7)

$$R_0 = R[T_{op,1}] \mid R[l] \le R[T_{op,1}], \; 40 \le l, T_{op,1} \le 79$$ (6.8)

$$R_0 = R[T_{op,2}] \mid R[l] \le R[T_{op,2}], \; 80 \le l, T_{op,2} \le 149.$$ (6.9)

The retained peak autocorrelation values are normalized according to [41,51]

$$r_i = \frac{R_i}{\sqrt{\sum_n u^2[n - T_{op,i}]}}; \; i = 0,1,2.$$ (6.10)

An overall winner is selected among these three outcomes by favoring the delays in the lowest range. The decision process is summarized as follows:

$$T_{op} \leftarrow T_{op,0}; r \leftarrow r_0;$$
$$\text{if } r_1 > 0.85r$$
$$T_{op} \leftarrow T_{op,1}; r \leftarrow r_1;$$
$$\text{if } r_2 > 0.85r$$
$$T_{op} \leftarrow T_{op,2}.$$

Thus, the procedure divides the lag range into three regions, with the final result (T_{op}) inclining toward the lower region, which is done to avoid choosing pitch multiples [41,51].

6.1.2.3 Search Range for Pitch Period

The pitch period of the adaptive codebook index is encoded with 8 bits for the first subframe. A relative value with respect to the first subframe is encoded with 5 bits for the second subframe. In the first subframe, a fractional pitch period with a resolution of $\frac{1}{3}$ is used in the range of $[19\frac{1}{3}, 84\frac{2}{3}]$, and the integral pitch period is used in the range of $[85,143]$. For the second subframe, a delay with a resolution of $\frac{1}{3}$ is always used in the range $[41,51]$

$$T_2 \in \left[round(T_1) - 5\frac{2}{3}, round(T_1) + 4\frac{2}{3} \right],$$

where T_1 and T_2 are the pitch periods of the first and second subframes respectively. For each subframe, the pitch period is determined by searching through the adaptive codebook in such a way that the weighted mean-squared error is minimized. The search range is found from the open-loop estimate T_{op}, which is roughly $T_{op} \pm 3$ and $T_{op} \pm 5$ for the first and second subframes respectively.

6.1.2.4 Finding Integer Lag

Given the preliminary code vector $d2'^{(t)}[n]$, it is processed by the filter cascade to generate the output sequence $y2'^{(t)}[n]$; that is [41,51]

$$y2'^{(t)}[n] = d2'^{(t)}[n] * h[n] = \sum_{k=0}^{n} h[k]d2'^{(t)}[n-k] \qquad (6.11)$$

for $t = T_{min} - 4$ to $T_{max} + 4$ and $n = 0$ to $N - 1$, with T_{min} and T_{max} found in the previous step. Note that the range of t has been extended; this is done to enable interpolation. Normalized correlation is calculated by using [41,51]

$$R[t] = \frac{\sum_{n=0}^{39} u3[n]y2'^{(t)}[n]}{\sqrt{\sum_{n=0}^{39} \left(y2'^{(t)}[n]\right)^2}}. \qquad (6.12)$$

The value of t that maximizes $R[t]$ within the interval $[T_{min}, T_{max}]$ is denoted as T, which is the integer lag sought.

6.1.2.5 Finding Fractional Lag

If it is the first subframe and the integer lag satisfies $T > 84$, the fractional part of the pitch period is set to zero. Otherwise, the fractions around T are evaluated by interpolating the correlation values of Eq. (6.12) in the following way [41,51]:

$$R_0(t, f) = \sum_{i=0}^{3} R[t-i]w13[3f + 3i] + \sum_{i=0}^{3} R[t+i+1]w13[3 - 3f + 3i] \qquad (6.13)$$

for $f = 0, \frac{1}{3}$, and $\frac{2}{3}$. In actual implementation, five fractional values are considered:

$$f = -\frac{2}{3}, -\frac{1}{3}, 0, \frac{1}{3}, \text{and } \frac{2}{3}.$$

The G.729 adopts such a strategy for the adaptive codebook search that is intended to be computationally efficient with some sacrifice on overall accuracy [51], especially for short pitch periods $(t + f < 40)$. It can be concluded that the approach is suboptimal in the sense that the best possible solution is not found, with the benefit of reducing computational burden. In practice, however, the method in combination with other components of the coder produces synthetic speech of high quality.

6.1.3 G.729 CODING SCHEME

The G.729 follows the basic schemes for encoding and decoding as for a generic CELP coder (Chapter 5). Differences exist mainly because of the extra complexity introduced to achieve higher performance. ACELP follows the fundamental structure of CELP, with a fixed codebook design that allows fast search with no storage requirement. Particularly for G.729, various innovations in the form of refined structure with added complexity are incorporated [51]. The most distinguishing features use the algebraic excitation codebook and conjugate VQ for the involved gains. The extra complexities essentially reflect the progress in speech coding development, where more sophistication is added to the coder framework for general improvement.

6.1.3.1 Encoding Operations

Figure 6.3 shows the block diagram of the G.729 encoder [41,51].

The input speech signal is high pass filtered and segmented into 80-sample frames (10 ms), subdivided into two 40-sample subframes. LP analysis is performed once per frame. Two sets of original and quantized LPCs are obtained at the end. Both sets are interpolated so as to apply to each individual subframe. The original LPCs (after interpolation) are used by the perceptual weighting filter. The input speech subframe is then filtered by it, with the output used in the open-loop pitch period estimation.

The impulse response of the filter cascades between the perceptual weighting filter and the formant synthesis filter is calculated. The resultant filter with system function $W(z) / \hat{A}(z)$ is referred to as the weighted synthesis filter. Note that the formant synthesis filter utilizes the quantized LPCs $\hat{a}_i$, with a system function [41,51]

$$\frac{1}{\hat{A}} = \frac{1}{1 + \sum_{i=1}^{10} \hat{a}_i z^{-i}}. \tag{6.14}$$

Table 6.1 [41,51] summarizes the bit allocation scheme of the G.729 coder. A total of 80 bits are allocated per frame, leading to a bit-rate of 8000 bps. One parity bit is transmitted per frame for error detection.

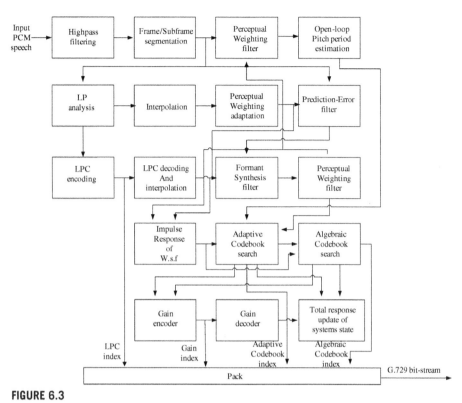

FIGURE 6.3

Block diagram of the G.729 encoder (w.s.f. = weighted synthesis filter).

Table 6.1 Bit Allocation for the G.729 Coder

Parameter	Number per frame	Resolution	Total bits per frame
LPC index	1	18	18
Pitch period (adaptive codebook index)	2	8, 5	13
Parity bit for pitch period	1	1	1
Algebraic codebook index	2	17	34
Gain index	2	7	14
Total			80

6.1.3.2 Decoding Operation

Figure 6.4 [41,51] shows the G.729 decoder.

The algebraic code vector is found from the received index, specifying positions and signs of the four pulses, which is filtered by the pitch synthesis filter in zero state. Parameters of this filter are specified in the previous section. The adaptive code vector is recovered from the integer part of the pitch period and interpolated according to

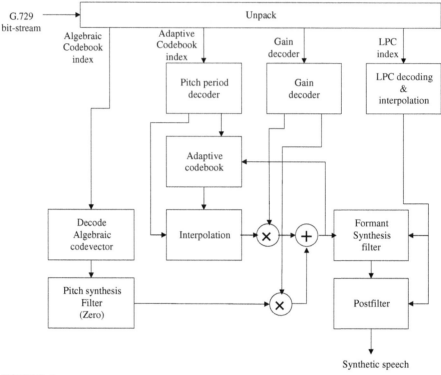

FIGURE 6.4

Block diagram of the G.729 decoder.

the fractional part of the pitch period. Algebraic and adaptive code vectors are scaled by decoded gains and added to form the excitation for a formant synthesis filter. A postfilter is incorporated to improve subjective equality [41,51].

6.2 THE ACELP-BASED SCHEME OF SPEECH INFORMATION HIDING AND EXTRACTION

The approach of ABS speech information hiding and extraction based on the G.729 scheme is realized as follows.

6.2.1 EMBEDDING SCHEME

The realization of the LCELP-based speech information embedding scheme is shown in Figure 6.5 [89].

In this scheme, carrier speech $x(n)$ is a 16-bit uniformly quantized PCM signal, with a sample rate of 8 kHz. This PCM signal is divided into frames, each consisting of 80 sample point values [39,136].

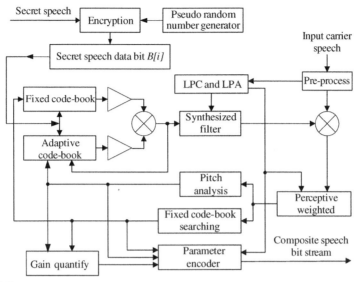

FIGURE 6.5

Scheme of ACELP-based speech information embedding.

A speech coder outputs coding parameters, which are closely correlated with the secret speech information data bits that should be embedded into carrier speech. These parameters are as follows [89]:

- LPF (Linear Predictive Filter) coefficient: 18 bits after being converted into LSP (Linear Spectral Pair) parameters
- Adaptive codebook ID: 13 bits
- Fixed codebook ID: 26 bits
- Adaptive code vector gain: combined with fixed code vector gain, for a total of 14 bits
- Fixed code vector gain: combined with adaptive code vector gain, for a total of 14 bits
- Pitch delay parity check parameters: 1 bit.

The adaptive codebook ID is composed of two parts: one is 8 bits, used for the first subframe, and the other is 5 bits, used for the second subframe. The fixed codebook ID consists of two parts, and each one is 13 bits corresponding to one subframe. Indeed, the codebook ID is the lookup address.

Not all the parameters can be used for secret speech information hiding. Experiments show that a slight modification to LSP may result in great harm to the speech quality. This is because a modification to the LSP coefficient will lead to the change of all sampled points that pass through the linear predictor. The change is dependent on the input sampled points, which cannot be estimated beforehand. Especially during the period of transition, when speech intensity is changed from weak to strong or from strong to weak, it will lead to a greater change of sample points. In fact, the period of transition is a very important part in speech, carrying more information. It may bring great change to some sampled points, which makes a big difference between original speech and synthetic speech in speech quality. This is similar to the parameters of adaptive code vector gain and fixed code vector gain and the fluctuation of their values, which greatly affects synthetic speech quality, and introduces more noise into synthetic speech. Hence, these parameters cannot be used for secret information hiding [89].

The adaptive codebook in the G.729 scheme adopts a fraction delay of 1/3 resolutions. The smallest change affects the adaptive codebook, which is 1/3 sampled points. The pitch frequency ranges from 50 Hz to 400 Hz. Pitch error $g(x)$ produced by the change of the adaptive codebook is expressed in the following equation [41]:

$$g(x) = \left(8000\Big/x\right) - \left(8000\Big/\left(x + \tfrac{1}{3}\right)\right) \quad x = 20, 160, \tag{6.15}$$

where x is the pitch period.

When $x = 20$, maximum $g(x)$ is equal to $400 - 8000\Big/\left(20 + \tfrac{1}{3}\right) = 6.557$ Hz. When $x = 160$, maximum $g(x)$ is equal to $50 - 8000\Big/\left(160 + \tfrac{1}{3}\right) = 0.104$ Hz.

This calculated result shows that the pitch error $g(x)$ introduced by slightly modifying the ID of the adaptive codebook is in proportion to the pitch frequency, which has a slight effect on synthetic speech, and even less of an effect on speech, which has lower frequency pitch. Therefore, the ID of the adaptive codebook is one parameter that can be used for information hiding. The result also suggests that the carrier speech must be carefully selected to achieve optimal information hiding when adopting the ID of an adaptive codebook as the embedding parameter in the proposed ABS information hiding approach. For example, the carrier speech selects the man's speech, whose pitch frequency is lower. Another parameter selected as the embedding parameter in the proposed ABS information hiding approach is the ID of a fixed codebook. Generally, the ID of a fixed codebook is better than the ID of an adaptive codebook in the quality of synthetic speech [41,89].

Based on this analysis and experiment, the ID of an adaptive codebook and the ID of a fixed codebook are determined as the embedding parameters used for secret information hiding in the proposed approach. They are changeable speech coding parameters.

According to the G.729 speech coding scheme, one frame of input data is composed of two parts. The first part is 80 sampled points $S(n)$ of the carrier speech, and the other is a group of 8 bits of secret information $B[i]$, $i = 0,\dots,7$.

In the following depiction, the implementation of embedding 8 bits into one frame and extracting 8 bits of embedded data from each frame shows the embedding process [41,89].

1. Analyze every sampled point in each frame using LP, convert them to an LSF (Linear Spectrum Frequency) parameter, then vector quantize for coding them into 18 bits.
2. Divide each frame into subframes of 40 sampled points.
3. Construct a perceptive weighted filter $W(z)$ and synthesis filter $H(z)$ individually by use of quantitative and nonquantitative LP parameters.
4. Select the adaptive codebook and fixed codebook under the control of embedding data sets $B[i]$ and predetermined constraint function F.
5. Multiply the two selected exciting signals by their own gains individually, adding the two results together to get a new exciting signal as the input to synthesis filter $H(z)$, which generates a local reconstruct signal $S(n)$. (A fixed codebook is generated by its algebraic construction, whose selection control depends on whether it meets the selection requirements. The gain of the adaptive codebook and fixed codebook is vector quantities in conjugating construction.)
6. Obtain the optimal exciting signal, which determines the MSE minimum by calculating the perceptive weighted MSE between $S(n)$ and $\tilde{S}(n)$.
7. Output the codeword of the determined optimal exciting signal in the current frame from the embedding encoder. This codeword, named CW, consists of the signal's ID in the codebook, gain, LP parameters, and other parameters. The embedding operation of this frame ends.

Here, function F is on sets, which makes the one-to-one relationship between CW and $B[i]$

$$B[i] \xrightleftharpoons[F^{-1}]{} CW,$$

where CW is an output codeword generated by an embedding coder used for the current frame and $B[i]$ are embedded data bits.

Carrier speech $x(n)$ is a uniform quantity PCM signal and a subframe (vector) that consists of five sample values. There are three codebooks: TABLE0, TABLE1, and TABLE-Z. TABLE0 and TABLE1 consist of 64 dependent code vectors individually; TABLE-Z consists of eight unit values that have zero symmetry, one of them being a sign bit, and two bits amplitude. When inputting a vector, the embedding coder selects a code vector in TABLE0, TABLE1, and TABLE-Z in sequence:

- If the bit value of the embedded data is 0, the vector in TABLE0 and TABLE-Z is selected.
- If the bit value of the embedded data is 1, the vector in TABLE1 is selected.

The select vector is amplified by a gain controller as an exciter of the synthesis filter, which generates a local decoded signal. Calculating the frequency weighted *MSE* between the local signal and the original signal, the result can be obtained by [41,51]

$$MSE = \|x(n) - \tilde{x}_{ij}\| = \sigma^2(n)\|\hat{x}(n) - g_i H y_j\|, \tag{6.16}$$

where H is the response function of the synthesis filter cascade with perceptual weighted filter; g_i is the i th gain; y_j is the j th code vector in TABLE0 or TABLE1; and $\hat{x}(n) = x(n) / \sigma(n)$ [88,89].

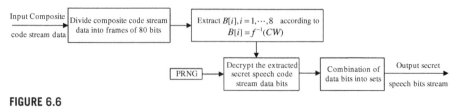

FIGURE 6.6

Scheme of ACELP-based speech information extraction.

The code vector coordinating to the minimum MSE is an optimal code vector, which is signed by a 10-bit sign and output as a result of embedding code. A frame consists of four optimal vectors, and each frame has a new LP factor. Exciting gain is obtained based on previous vectors and updated with each new vector.

6.2.2 EXTRACTION SCHEME

The realization of an LCELP-based speech information extraction scheme is shown in Figure 6.6 [89].

When the secret message is embedded by the embedded coder, it outputs codeword CW, which satisfies the mapping relationship F of $B[i]$ and CW. At the receiving end, the decoder can extract the embedded secret message once it has received the composite carrier data coding stream, without any information about original carrier speech.

In the scheme of ACELP-based speech information extraction, speech data are divided into frames of 80 bits. The extraction procedure follows:

1. Divide the received composite carrier coding data bit-stream into groups of 10 bits, which is composed of a codeword CW.
2. Calculate the secret message data bits value $B[i]$ by using the inverse mapping relationship F^{-1}, i.e. $B[i] = F^{-1}$.
3. Decrypt, and recover the coding of the original secret message data bits $B[i]$, and output the data bit-stream corresponding to the coding order of original secret message data bits $B[i]$ to the secret speech information decoder.
4. Synthesize secret speech $i = 1, \cdots, 8$.

6.3 APPROACH TO HIDE SECRET SPEECH IN G.729

This section proposes an approach of speech information hiding based on the G.729 speech coding scheme [51], extending information hiding application to real-time speech secure communication.

6.3.1 EMBEDDING ALGORITHM

An ABS embedding algorithm model (Figure 6.7) is based on the analysis of parameter features of the G.729 scheme, and constructs an information embedding model [85,89].

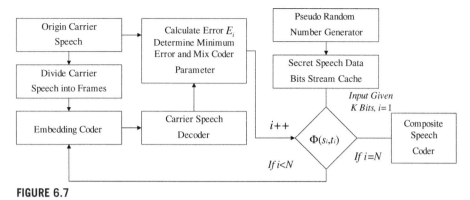

FIGURE 6.7

Embedding algorithm.

To better understand the embedding algorithm model, the notations used in analysis are listed as follows. They are definite in the $\mathbb{Z}$ field and $\mathbb{Z}$ includes components of N [89].

M	A speech information data frame length, sample rate of 8 kbps
c_j	Carrier speech code stream
s_i	A frame of secret speech code stream
$\Phi(s_i,t_i)$	Embed method (substitute method, or transform)
t_i	i th frame
$t(i)$	The original speech sample point of frame t_i
$t'(i)$	Sample point of composite speech
E_i	The error value between original speech and mixed speech
CW	Speech codeword; it can be a quantity or a vector, in which many parameters are involved
BV	The embedded speech information data bit value
f	The function expressing the relationship of CW and BV: $BV = f(CW)$

The equation $BV = f(CW)$ means that various speech information data bits are embedded into various parameters and various positions. The ABS embedding algorithm involves three steps [89]:

1. Divide the original speech into frames. To meet the requirements of secret speech information data hiding, the length of the speech frames is not the same according to the speech coding standard and the feature of the speech coding algorithm.
2. Embed speech coding to adopt the G.729 standard, and the parameters after coding are satisfied with $BV = f(CW)$; then feed back these parameters to an analysis-synthesis system. For each $\Phi(s_i,t_i)$, mixed speech parameters, it must be calculated and decoded to original speech. Error is calculated according to the equation:

$$E_i = \sum_{i=1}^{M}[t'(t) - t(i)]^2, \quad i = 1, 2, \cdots, N. \tag{6.17}$$

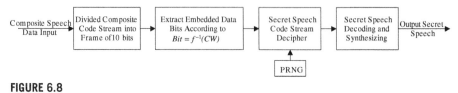

FIGURE 6.8

Extraction algorithm.

3. Calculate $E_{min} = \min\{E_i; i = 1, 2, \cdots, N\}$, determine $\Phi(s_{min}, t_{min})$, and then output the *CW* coordinates with the method $\Phi(s_{min}, t_{min})$.

6.3.2 EXTRACTION ALGORITHM

The G.729-based secret speech extraction algorithm is shown in Figure 6.8 [89].

The received composite speech code stream data are divided into sets by 10 bits; that is, a codeword *CW*. According to the relationship between codeword *CW* and data bit value *Bit*, the embedded data bit value *Bit* can be obtained by using the following equation:

$$Bit = f^{-1}(CW). \qquad (6.18)$$

The G.729-based secret speech extraction algorithm is speedy and concise because it does not need the original carrier speech.

6.4 EXPERIMENTAL RESULTS AND ANALYSIS

Two sections of speech are recorded, the longer one for the original speech carrier and the shorter one for secret speech. The original carrier may be used in a cycle in order to embed the entire secret speech into it. The original speech is divided into small sections of 20 ms, consisting of four frames of 5 ms each.

In our proposed approach, experiments adopt G.729 as the speech coding scheme of carrier speech. Secret speech information data hiding capacity is 800 bps, 2400 bps, and 3200 bps, individually. Original carrier speech is randomly selected, and secret speech is the phrase "Welcome to the information hiding world" spoken by a middle-aged man [51].

The analysis of the test results is based on the spectrum coming from the original carrier speech, composite speech, and extracted secret speech in G.729. Comparing the spectrum of original carrier speech (Figure 6.9) with that of composite speech (Figure 6.11) [89], analysis shows that our proposed approach keeps the speech's continuity and understandability, and has a high capacity and real time. Comparing the spectrum of original secret speech (Figure 6.10) with that of extracted secret speech (Figure 6.12) [89], analysis shows that they are slightly different in signal amplitude and have small delay, but these defects hardly affect the understanding of the speech. The result of spectrum comparison is satisfactory, corresponding with actual hearing.

FIGURE 6.9

Original carrier speech spectrum.

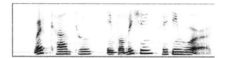

FIGURE 6.10

Original secret speech spectrum.

FIGURE 6.11

Composite speech spectrum (G.729).

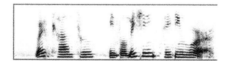

FIGURE 6.12

Extracted secret speech spectrum.

6.5 SUMMARY

In the design and implementation of secure communication based on the technology of speech information hiding, the public/carrier speech coding scheme is selected depending on the requirement of the communication channel capacity. The maximum hiding capacity of the proposed ABS speech information hiding approach can be estimated under the conditions of fixing the carrier speech coding scheme and setting the constraints of hiding effectiveness. The selection of the error measurement method and the subjective evaluation of composite speech quality are two constraint conditions to the maximum hiding capacity estimation. If information hiding requirements such as capacity, robustness, and invisibility are satisfactory, more constraint conditions should be presented to determine the method of secret information extraction and make the algorithm of embedding and extraction simpler, making the practical speech information hiding system more efficient.

In this chapter, an approach of secure communication based on G.729 is presented. In this approach, speech coded in the G.729 scheme is used as public/carrier speech and a 2.4 kbps MELP speech is selected as secret speech. The proposed approach realizes the secure communication at 16.0 kbps with secret speech being embedded at the rate of 0.8 kbps using the LPC-IH-FS algorithm. The secret speech information is extracted using the BD-IE-MES algorithm. The public speech coded in the G.729 scheme is used to build two common communication channels, an open channel used for public speech transmission and a subliminal channel used for secret speech communication. The proposed approach uses the redundancy to the greatest extent under different constraint conditions and various requirements; it is a secure and safe communication technique for secret speech information.

Test results show that the information hiding capacity of our approach can reach 800 bps with good robustness if an 8 kbps G.729 scheme is used in our proposed approach.

Adopting an ABS algorithm will improve the speech quality when secret speech information data bits are embedded. The secure communication device, SIHT, is developed based on the presented approach for secure communication via PSTN. SIHT is tested within a PSTN environment, and the results show that SIHT can communicate securely at 8 kbps, which has been achieved in more than 94% of the trials, with excellent speech quality and complicating speakers' recognition, even though understanding is guaranteed.

The GSM (RPE-LTP)-Based Speech Information Hiding Approach

7

GSM is short for Full Rate (FR); it is sometimes expressed as GSM-FR or GSM 06.10, which is the first digital speech coding standard used in the GSM digital mobile phone system. Although the quality of GSM's coded speech is quite poor by modern standards, the codec is still widely used in networks around the world. GSM-FR is based on the Regular Pulse Excitation–Long Term Prediction (RPE-LTP) speech coding paradigm. The RPE-LTP speech coding scheme has two advantages. On one hand, it has a relatively low coding bit rate of 13 kbps and leads to a good quality of reconstruction speech. On the other hand, some parameters of the coding bit have better robustness and thus a few changes make no sense for the reconstruction speech [137].

This chapter applies the idea of ABS-based speech information hiding to the RPE-LTP speech coding scheme to realize GSM-based secure communication. ABS-based speech information hiding and extraction algorithms are applied to the GSM based on the approach that was proposed in Chapter 3. In this chapter, a brief introduction to the GSM speech coding scheme is given first and then the detailed embedding and extraction operations are introduced.

7.1 INTRODUCTION TO THE GSM (RPE-LTP) CODING STANDARD

The GSM speech encoder accepts 13-bit linear PCM at an 8 kHz sample rate; that is, the bit rate of the codec is 13 kbit/s [138]. This can be directly obtained from an analog-to-digital converter in a phone or computer, or converted from a G.711 8-bit nonlinear A-law or μ-law PCM from the PSTN with a lookup table. In GSM, the codec operates on 160 sample frames that span 20 ms, so this is the minimum possible delay of a transcoder even with infinitely fast CPUs and zero network latency. Like many other speech codecs, linear prediction is used in the GSM synthesis filter. The order of the linear prediction is only 8. In modern narrowband speech codecs the order is usually 10, and in wideband speech codecs the order is usually 16 [41].

Information Hiding in Speech Signals for Secure Communication. DOI: 10.1016/B978-0-12-801328-1.00007-0

7.1.1 **RPE-LTP CODING SCHEME**

Within the context of LP, prediction error is obtained by filtering the speech signal using the prediction-error filter. The original signal can be reconstructed by passing the prediction-error sequence through the synthesis filter. The synthesis filter and prediction-error filter are the inverse of each other. Using LP (with the coefficients updated on a frame-by-frame basis), an ADPCM system can be designed by quantizing the samples of the prediction error. In fact, many prediction-error samples have relatively low amplitudes, and it is not necessary to encode all prediction-error samples in order to achieve good quality for the reconstructed speech. Typically, by preserving only 10% of the samples (the other 90% are set to 0), different kinds of speech sounds can be generated with little perceptual distortion. The resultant approach is called the multipulse excitation model since the excitation sequence is formed by scattered pulses, in which most of the samples are set to zero [41,122,135].

7.1.1.1 *Multipulse Excitation Model*

As described in Chapter 4, the predictor of the ADPCM system is made adaptive. The method is practical, but it is highly redundant that unnecessary information is being sent as part of the bit-stream, needlessly increasing the resultant bit-rate. The main source of redundancy is from the prediction-error signal itself. Since many samples are low in amplitudes, their elimination from the bit-stream causes diminutive perceptual distortion [50,85].

The multipulse excitation model is based on the observation that only a small percentage of nonzero samples are necessary to excite the synthesis filter so as to produce good quality speech. The nonzero samples are encoded by their amplitudes and positions. The problem is therefore finding a set of pulses that minimizes distortion. The number of pulses per frame is directly related to bit-rate and quality and must be selected to satisfy the constraints to an application. Many methods have been proposed to solve the problem of multipulse excitation. In general, these methods can be classified as *open loop* and *closed loop*.

Open-Loop Method

Figure 7.1 shows the block diagram of a coding system based on a multipulse model, with the pulses selected via an open-loop method [41,122,135]. The input speech is LP analyzed to obtain the LPCs, which are used to construct the prediction-error

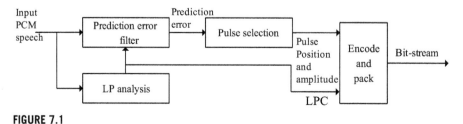

FIGURE 7.1

Encoder of an open-loop multipulse coder.

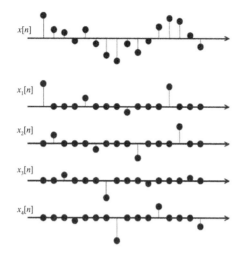

FIGURE 7.2

Illustration of regular-pulse excitation.

filter. The resultant prediction-error sequence is processed by the pulse selection block, where a certain criterion is applied to eliminate the majority of samples. The selected samples are encoded by their amplitudes and positions. The decoder recovers the information, regenerates the excitation sequence, and passes it through the synthesis filter to obtain the synthetic speech. Many criteria can be used in the pulse selection block.

In a simple scheme based on the magnitude of the pulses, the samples are first sorted accordingly, and only a fixed number of the highest magnitude samples are retained. This approach, however, requires the coder to take note of the amplitude as well as the position of every selected pulse. One popular method is the regular-pulse excitation scheme, illustrated in Figure 7.2 [41,122,135]. In this approach, the prediction-error sequence is down-sampled to a number of different sequences, with the highest energy sequence selected. In this way, only the amplitude of the pulses and the sequence number need to be encoded, reducing the number of bits carrying information.

Closed-Loop Method

Figure 7.3 shows the encoder of a closed-loop system [41,122,135].

In this method, the excitation sequence is selected through a feedback loop, where the difference between input speech and synthetic speech is minimized. Note that speech synthesis is actually performed during encoding; hence, the method is also known as ABS, due to the fact that the signal is analyzed by synthesizing a replica for comparison. Within the loop, parameters of the signal are determined so as to minimize error. In the present case, the parameters are the amplitudes and positions of the pulses that constitute the excitation sequence.

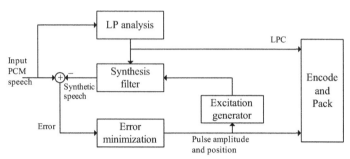

FIGURE 7.3

Encoder of a closed-loop multipulse coder.

7.1.1.2 Characteristics of GSM (RPE-LTP) Coding Scheme

The GSM coding scheme has a relatively low bit rate (13 kbps) [135,137], and the reconstruction of the synthetic speech retains good quality. Some parameters used here have a relatively strong robustness. Therefore, a small change to these parameters does not greatly influence the quality of reconstructed speech.

7.1.2 GSM CODING SCHEME

The encoder of GSM 8.10 RPE-LTP is essentially an ADPCM system, where a predictor is computed from the signal, with the prediction error found, and is subsequently quantized using an adaptive scheme. The predictor is implemented as a cascade connection of short-term and long-term predictors. Long-term predictors greatly increase the average prediction gain, thus elevating the overall performance [138].

Figure 7.4 shows the block diagram of the encoder, where parameters of each frame/subframe are extracted and packed to form a bit-stream. Each frame has a length of 160 samples (20 ms) and is subdivided into four subframes of 40 samples each [41,122,135].

7.1.2.1 Preprocessing

The preprocess filter with system function $H(z) = (1 - z^{-1}) / (1 - 0.999z^{-1})$ is used as a high-pass filter to eliminate any zero-frequency (DC) component. The resultant signal is preemphasized using a filter with system function $1 - 0.86z^{-1}$.

7.1.2.2 LP Analysis

LP analysis is performed on every 160-sample frame. An eight-order prediction is employed. Nine autocorrelation values are calculated from the frame by using a rectangular window. The autocorrelation values are solved to obtain eight reflection coefficients (RCs).

7.1.2.3 LPC Quantization and Interpolation

The RCs are transformed to LARs (logarithm area ratios) through a piecewise linear function and are then scalar quantized by using 36 bits. An index of the quantized LPCs is transmitted as part of the bit-stream. The quantized LARs are interpolated to four different sets to be used by the four 40-sample subframes.

7.1.2.4 Prediction-Error Filter

The interpolated LARs are transformed back to RCs, which are used to form the prediction-error filter. This filter is implemented in lattice form. The filter produces the (short-term) prediction-error signal (internal prediction) at its output.

7.1.2.5 Long-Term LP Analysis, Filtering, and Coding

The "Long-term LP analysis, filtering, and coding" block in Figure 7.4 is redrawn in Figure 7.5 [41,122,135].

Note that the operations described next are repeated once every subframe, or every 40 samples. The short-term prediction error d from the (short-term) prediction-error filter together with the reconstructed prediction error d' are used for long-term LP analysis. The cross-correlation function [41,122,135] is

$$R(T) = \sum_n d(n)d'(n-T), \quad T = 40,\ldots,120. \tag{7.1}$$

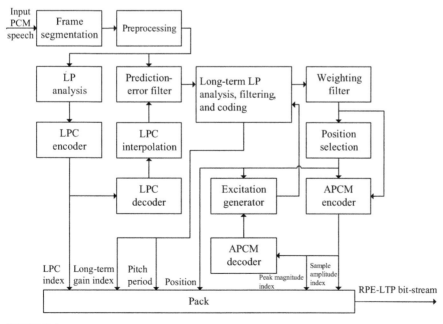

FIGURE 7.4

Block diagram of the RPE-LTP encoder.

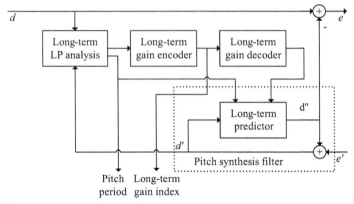

FIGURE 7.5

Implementation of the long-term LP analysis, filtering, and coding block.

That is, a total of 81 pitch period values are utilized, encoded with 7 bits. The range of n corresponds to the subframe under consideration, with 40 samples having been taken into account. Thus, the range of T is selected in such a way that the delayed samples (from d') used in the cross-correlation calculation are outside the current subframe. Pitch period of the subframe is defined as the one that maximizes Eq. (7.1) [41,122,135]. Once the pitch period of the subframe is found, the long-term gain is calculated as [41,122,135]

$$b = \frac{R(T)}{\sum_{n}[d'(n-T)]^2}.$$
(7.2)

Since d' is the actual sequence employed for speech synthesis, its usage yields higher performance due to the fact that both encoder and decoder can operate by using the same signal.

The pitch synthesis filter takes the reconstructed overall prediction error e' to generate the reconstructed short-term prediction error d' [41,122,135]

$$d'(n) = e'(n) + d''(n),$$
(7.3)

where d'' is the prediction given by [41,122,135]

$$d''(n) = -\hat{b}d'(n-T).$$
(7.4)

7.1.2.6 Weighting Filter

The overall prediction-error sequence of the current subframe is filtered by an 11-tap FIR filter. Plots of impulse response and frequency response are shown in Figure 7.6. The filter is low-pass in nature. Its incorporation smoothes out variations between samples, suppresses high-frequency noise, and makes transitions

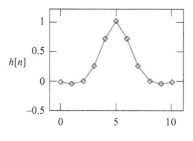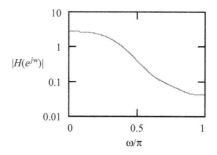

FIGURE 7.6

Impulse response and frequency response of the weighting filter.

between subframes smoother, thus improving the subjective quality of the synthetic speech [41,122,135].

7.1.2.7 Position Selection

The filtered prediction-error sequence is down-sampled by a ratio of 3, resulting in four interleaved sequences with regularly spaced pulses. These are defined with [41,122,135]

$$x_m(n) = x(m+3n); \quad m=0,1,2,3; \quad n=0,1,\ldots 12. \tag{7.5}$$

This equation defines four sampling grids, illustrated in Figure 7.7 [41,122,135].

The best candidate sequence $x_m(n)$ is selected by verifying the energy [41,122,135]

$$E_m = \sum_{n=0}^{12} x_m^2(n) \quad m=0,1,2,3 \tag{7.6}$$

with the one producing the highest energy selected. The resultant grid position m is coded with 2 bits.

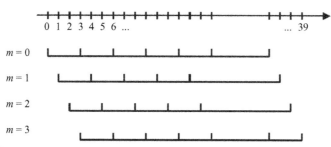

FIGURE 7.7

Sampling grids used in position selection.

7.1.2.8 APCM

The selected subsequence $x_m(n)$ is quantized using a forward gain-adaptive quantizer. For each 13-sample sequence, the peak magnitude is determined by $g = \max|x_m(n)|$, which is quantized by using a nonlinear (logarithmic) quantizer with 6 bits. The quantized peak magnitude is used to normalize the samples $x'_m(n) = x_m(n)/\hat{g}$, and are quantized by using a 3-bit uniform quantizer [41,122,135].

7.1.2.9 Excitation Generator

The quantized samples of the normalized sequence are denormalized with the quantized peak magnitude. The resultant sequence is up-sampled by a ratio of 3 by inserting zero samples according to the grid position. The final signal is denoted by e', as shown in Figure 7.5.

7.2 APPROACH TO HIDE SECRET SPEECH IN GSM (RPE-LTP)

In order to describe the embedding and extraction algorithm better, the variable symbols used in the algorithms are presented here [25] :

T_1 is the time span of coding one frame.

T_2 is the time span of employed one frame low-rate codes.

T refers to the frame length of the embedding algorithm (or the length of buffer).

$T = [T_1, T_2]$ is the least common multiple of T_1 and T_2. To ensure that the carrier and secret speech can be sent synchronously, the delay of embedding T should be 180 ms.

f expresses the relationship between embedded bits and encoded codewords, and f^{-1} represents the inverse relationship.

7.2.1 EMBEDDING ALGORITHM

The approach to hide secret speech in GSM coding is shown in Figure 7.8 [25,41]. The embedding procedure is as follows [25]:

1. The carrier speech is preprocessed by the sender to eliminate the DC component and preemphasize the high-frequency components in order to get better LPC analysis. The preemphasizing filter adopts 1-order FIR filter.
2. The speech is divided into frames after signal segmented preprocessing. Each frame consists of 160 samples (20 ms).
3. The short-term redundant signal d is generated by performing LPC short-term prediction analysis to signal S.
4. The redundancies in the redundant signal are eliminated by sending the redundant signal d through the long-term predictor after the short-term prediction analysis.

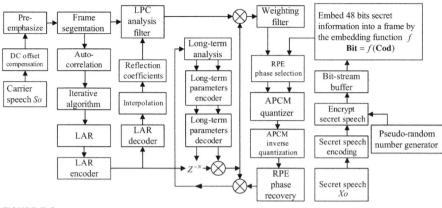

FIGURE 7.8

Block diagram of embedding algorithm.

5. *X* is the original secret speech that has been sampled after A/D transformation. It is generated by the selected low bit-rate encoder. X_i is the bit-stream of the *i* th secret bit frame (48 bits).

6. The codewords of composite speech are generated by embedding X_i into the carrier speech according to the predestinated algorithm.

7. The codewords of composite speech are output and will be transported by a transmission protocol.

7.2.2 EXTRACTION ALGORITHM

The flow chart of the extraction algorithm based on the GSM (RPE-LTP) coding scheme is shown in Figure 7.9 [25,41]. The extraction algorithm procedure is as follows [25]:

1. The composite speech bit-stream S_x is analyzed and preprocessed in order to extract the secret speech.

2. The composite speech bit-stream S_x is divided into a segment with a length of *T*.

3. The secret speech bit-stream X_i is extracted from S_{xi} according to the extraction algorithm by using the extraction function f^{-1}.

4. The extracted secret speech bit-stream X_i is decrypted by using the decryption algorithm.

5. The decrypted secret speech bit-stream is coded by using the selected low-rate speech coding scheme to generate the synthetic secret speech.

6. The synthetic secret speech is output to complete the secret speech information transmission.

In the flow chart illustrated in Figure 7.9, the composite speech is coded in GSM and output from another bypass for the wiretappers.

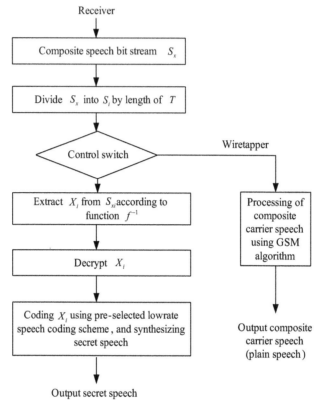

FIGURE 7.9

Flow chart of extraction.

The proposed extraction algorithm can extract the secret speech in a blind detection way. The composite bit stream should be divided into segments with a length of $T = 20$ ms. Secret speech information data in each segment is extracted depending on the information extraction function f^{-1}, which is similar to previous chapters.

7.3 EXPERIMENTAL RESULTS AND ANALYSIS

The hiding effects of the GSM (RPE-LTP) coding scheme are shown in Figures 7.10 through 7.17 [25], and the explanations are labeled in Table 7.1.

The experimental results show that the extracted secret speech has no distinct difference from the original secret speech, so the quality of the original secret speech was retained well.

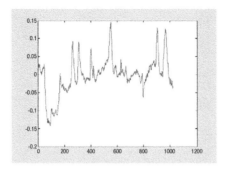

FIGURE 7.10

Original secret speech wave.

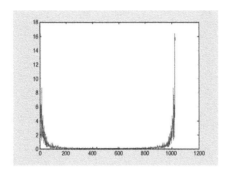

FIGURE 7.11

Original secret speech spectrum.

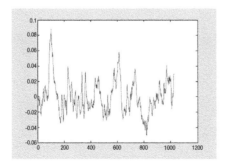

FIGURE 7.12

Original carrier speech wave.

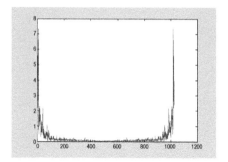

FIGURE 7.13

Original carrier speech spectrum.

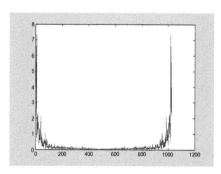

FIGURE 7.14

Composite speech wave.

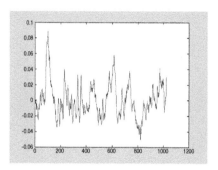

FIGURE 7.15

Composite speech spectrum.

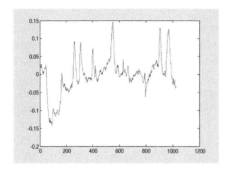

FIGURE 7.16

Extracted secret speech wave.

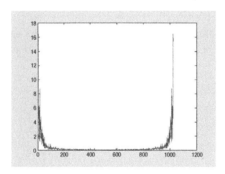

FIGURE 7.17

Extracted secret speech spectrum.

Table 7.1   Test Result of Effect and Hiding Capacity

Test result / Code schemes	MOS	DRT	DMA	Ψ	Φ	R
GSM and MELP	3.5	90%	Satisfied	Maximum 2600 bps	2.92 dB	13 kbps

Note: Ψ is the hiding capacity. Φ is the difference between normal speech coding with synthesis speech embedded coding in dB. R is the communication data rate.

7.4 SUMMARY

In this chapter, an approach of secure communication based on the technology of information hiding is realized on concrete speech coding according to the model of information hiding. The GSM-based secure communication approach is presented. In this approach, speech coded in the GSM scheme is used as public/carrier speech and 2.4 kbps MELP speech is selected as secret speech. The proposed approach succeeds in conducting a secure communication at 13.0 kbps with a secret speech embed rate of 2.6 kbps based on the algorithm of LPC-IH-FS. The extraction of secret speech information is completed by using the algorithm of BD-IE-MES. In this approach, the public speech coded in GSM is a carrier that completes two functions: first, it builds a common communication channel for public information; second, it sets up a subliminal channel for secret speech.

The proposed approach is adopted by SIHT, which has been tested in a PSTN environment. Results show that the proposed approach has achieved more than 94% of the trials at 13 kbit/s, with an excellent speech quality and complicating speakers' recognition, even though understanding is guaranteed.

The proposed approach can be modified flexibly to satisfy different speech information hiding requirements and different speech coding schemes.

Covert Communication Based on the VoIP System

8

This chapter focuses on the secure communication over VoIP based on the technology of speech information hiding. An approach of embedding secret speech information into public speech for secure communication over VoIP is proposed. A 2.4 kbps MELP speech is used as secret speech, and G.729 coding speech is used as the public carrier. The proposed approach satisfies the requirements of real-time transmission with good imperceptibility and large hiding capacity.

8.1 INTRODUCTION TO THE VoIP-BASED COVERT COMMUNICATION SYSTEM

The G.729 source codec, which is used extensively in VoIP, is studied to analyze the capability of noise tolerance (CNT) of different speech coding parameters [5,7]. Analysis reveals that many parameters can be used for carrying secret data because they have less impact on reconstructing the speech signals. These parameters include a fixed codebook index, fixed codebook signs, and the second stage higher vector of the LSP quantizer. Based on the conclusion, an algorithm is proposed to embed a 2.4 kbps MELP speech into a G.729 coding speech by adapting the covering code and the interleaving technique [51,89,142].

8.1.1 INTRODUCTION TO THE VoIP SYSTEM

Voice over Internet Protocol (VoIP) commonly refers to communication protocols, technologies, methodologies, and transmission techniques involved in the delivery of voice communications and multimedia sessions over Internet Protocol (IP) networks, such as the Internet [5,7,139]. Other terms commonly associated with VoIP are IP telephony, Internet telephony, Voice over Broadband (VoBB), broadband telephony, IP communications, and broadband phone. Internet telephony refers to communications services—voice, fax, SMS, or voice-messaging applications that are transported via the Internet, rather than the Public Switched Telephone Network (PSTN). IP networking may possess some benefits, due to the packet switched working mechanism. It is the fundamental communication protocol for computer communication, and it offers solutions to merge voice and other data.

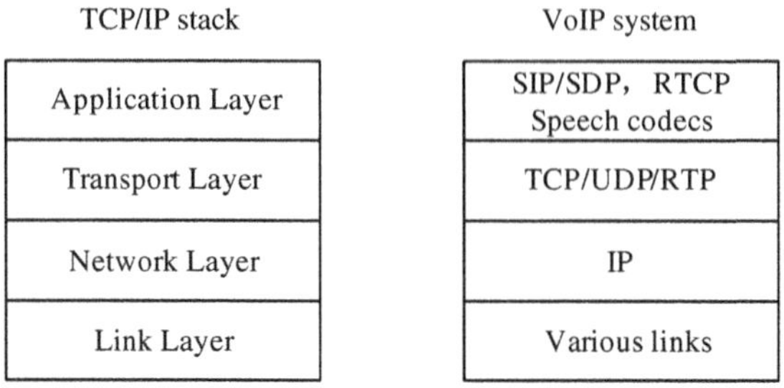

TCP/IP stack   VoIP system

Application Layer	SIP/SDP, RTCP Speech codecs
Transport Layer	TCP/UDP/RTP
Network Layer	IP
Link Layer	Various links

FIGURE 8.1

VoIP stacks and protocols.

The steps involved in originating a VoIP telephone call are signaling exchange and media channel setup process, digitization of the analog voice signal, encoding, packetization, and transmission as Internet Protocol (IP) packets over a packet-switched network [140]. On the receiving side, similar steps (usually in the reverse order) such as reception of IP packets, decoding of the packets, and digital-to-analog conversion may be followed to reconstruct the original voice stream.

VoIP systems employ session control protocols to control the set-up and teardown of calls as well as audio codecs, which encode speech allowing transmission over an IP network as digital audio via an audio stream. The choice of codec is performed according to implementations of VoIP depending on application requirements and network bandwidth. Some implementations rely on narrowband and compressed speech, while others support high fidelity stereo codecs. Popular codecs are μ-law and A-law versions of G.711 [66]: G.722, which is a high-fidelity codec marketed as HD Voice by Polycom; G.729 [141], which uses only 8 kbps each way; and many others [76,77,89,142,143]. Figure 8.1 shows VoIP stacks and protocols [144,145].

An IP telephony connection consists of two phases, signaling and conversation, in which certain types of traffic are exchanged between the calling parties. During the first phase, certain signaling protocol messages, for example SIP (Session Initiation Protocol) messages, are exchanged between the caller and callee. These messages are intended to set up and negotiate the connection parameters between the calling parties. During the second phase, two audio streams are sent bidirectionally. RTP (Real-Time Transport Protocol) is most often used for voice data transport, and the packets that carry the voice payload are called RTP packets. The consecutive RTP packets form an RTP stream [76,77].

8.1.2 AN OUTLINE FOR VoIP STEGANOGRAPHY

As mentioned in Chapters 1 and 2, steganography is an ancient technique aiming at embedding a secret message into a carrier. This method hides the very existence of the communication, and therefore keeps any third-party observers unaware of the presence of the steganographic exchange. Steganographic carriers have evolved through the ages and are related to the methods of communications, and a VoIP system is a perfect carrier for embedding for the following reasons [139]:

- It is very popular, and thus its utilization may not raise others' suspicions; that is, it will not be considered an anomaly.
- The volume of VoIP data is quite large; the more frequent the presence and usage of these carriers in networks, the bigger their embedding capacity [76].
- Potentially good embedding capacity may be achieved. For example, during the conversation phase of a G.711-based call, each RTP packet should carry voice data of 20 ms. In this circumstance, the RTP stream rate is 50 pps (packets per second). By simply hiding 1 bit in every RTP packet, a quite high embedding capacity of 50 bps can be achieved [145].
- Various protocols are used. As shown in Figure 8.1, a VoIP system involves a combination of many protocols. Thus many opportunities for embedding information arise from the different layers of the TCP/IP stack. Covert communication may be implemented by applying hiding methods to the users' voice data, which is carried inside the RTP packets' payload [77].
- Since secure communication is a real-time service, it induces additional strict requirements for synchronization and simultaneously creates new chances for embedding usage. For example, excessively delayed packets are discarded by the receiver because they are not useful for voice reconstruction.

8.1.3 CLASSIFICATIONS OF THE EMBEDDING METHOD

At present, embedding methods that can be used in telecommunication networks are jointly called network steganography, or specifically VoIP steganography when applied to VoIP. These terms are subjected to the techniques of hiding information in any layer of the TCP/IP protocol stack, as shown in Figure 8.1 [145], including some techniques applied to the speech codecs and those that use the network protocols themselves.

The expansion of TCP/IP networking has opened up many possibilities for covert communication. Because IP networking changes the traditional circuit-switched networks paradigm, services are created by users rather than the network itself, transport and controls are combined, and they can also be influenced by users. These features make it possible for users to influence and use the data control flow and the communication protocols together with service functions of terminals to set up covert communication data flows. This is why secret messages can be embedded, not only within ordinary overt messages as in traditional steganography and circuit-switched networks, but also in the communication protocols' control elements, in effect by manipulation of the protocols' logic, or by combinations of these features.

So far, a number of steganographic methods have been put forward, and they cover all the layers of the TCP/IP stack, as shown in Figure 8.1 [144].

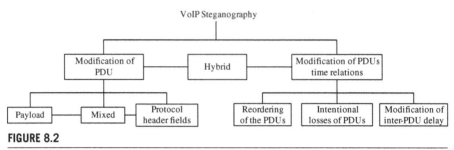

FIGURE 8.2

VoIP steganography classifications.

In addition, many of the previously proposed methods can be successfully applied to VoIP traffic. In general, based on the used carrier, embedding techniques can be classified into three groups (shown in Figure 8.2) [139]:

1. PDU (Protocol Data Unit)-based embedding methods—using network protocol headers or payload field. Examples of these solutions are [139]:
   • Modifying signaling messages in SIP data during signaling setup phase
   • Modifying redundant headers' fields of IP, UDP, or RTP protocols in conversation phase
   • Modifying RTP packets' payload by modifying the user content or simply by replacing users' data. According to the payload types of speech, the embedding approach introduced in Chapters 3 through 7 can be used in modifications of the RTP packets' payload.
2. Embedding methods that modify PDUs' time relations—by affecting the sequence order of PDUs, modifying PDUs' interpacket delay, or introducing intentional PDU losses. These approaches were not specifically proposed for VoIP systems, but they are suitable mainly due to the VoIP traffic characteristics and volume. Actually, time relations modification solutions are rather hard to deploy practically, because they generally provide low hiding capacity and require accurate synchronization information. Synchronization may take up a certain part of transmission bandwidth, and thus the hiding capacity decreases further.
3. Hybrid embedding methods modify both the content of PDUs and their time relations. These methods offer a new idea for embedding. For example, a large volume of secret data can be hidden by modifying the content of PDUs, while time relations modification can be used for secret keys exchanges.

8.2 MODELING AND REALIZATION OF VoIP-BASED COVERT COMMUNICATION

The model of a covert communication system based on IP telephony systems is depicted in Figure 8.3 [143,144].

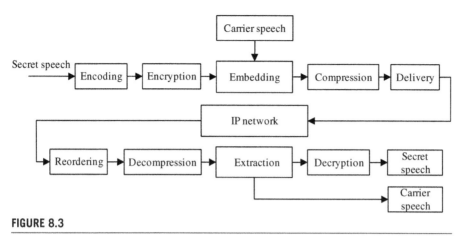

FIGURE 8.3

The model of covert communication based on the IP telephony system.

This system adopts information hiding techniques, and connects public IP networks like an ordinary IP telephone. Figure 8.3 also illustrates the one-way working procedure from sending end to receiving end.

This system takes real-time voice data as the carrier, and embeds secret data (i.e., secret speech) by using the redundancy of carrier speech. From the viewpoint of external behavior, two participants are conducting normal speech exchange, and secret speech is actually sent within these normal voice packets. Hence the goal of covert communication is achieved and secret speech has imperceptibility.

How to coordinate the relationship among hiding capacity, speech quality, and security should be considered when selecting an embedding approach. The hiding capacity must be large enough to meet the requirement of real time. The embedding process should not degrade the carrier speech too much, otherwise degradation may raise suspicion and attention from illegal users. In the rest of this chapter, an embedding method that maintains low process delay and keeps a good quality of carrier speech is discussed.

8.3 EMBEDDING SECRET SPEECH INTO VoIP G.729 SPEECH FLOWS

In this section, an embedding method will be introduced that differs from the methods introduced in the previous four chapters. By using matrix coding [146], hiding capacity may be high enough to meet the requirements of real-time communication, thus 2.4 kbps MELP coded speech can be embedded into 8 kbps G.729 coded speech. This method also utilizes the idea of least significant bit (LSB) [142]. In order to obtain the LSB in G.729 coded speech [141], experiments on Capability of Noise Tolerance (CNT) are completed first [85].

8.3.1 THE CNT OF G.729 PARAMETERS

From the viewpoint of the encoding principle, parameters are first obtained by linear prediction and quantization. Then the prediction parameters are adaptively adjusted by a feedback system in order to minimize the error between the original and reconstructed speech according to a perceptually weighted distortion measure. There is a CNT to a certain extent for these parameters, which means that the parameters allow a certain value of error.

By analyzing the principle of encoding, it is known that the linear prediction coefficients are converted to line spectrum pairs (LSPs) and quantized using predictive two-stage vector quantization (VQ). The first stage is a 10-dimensional VQ using a codebook L1 with 128 codewords (7 bits). The second stage is also a 10-dimensional VQ, which is divided into two 5-dimensional codebooks. L2 is lower vector and L3 is higher vector, including 32 codewords (5 bits). From the essence of the vector quantization, the line spectrum pairs are divided into different particle sizes in this process. The division will introduce errors of quantization for line spectrum pairs, but the second stage of quantization introduces fewer errors than the first stage. From this point, it is not difficult to see that the LSP parameter has a certain capability of noise tolerance. From the point of information hiding, it means that the parameter can be used for embedding the secret information [51].

The process of excitation production is that the adaptive codebook is searched in order to minimize the error between the synthetic signal and original signal. The fixed codebook is searched for minimizing the mean squared error between the weighted input speech and the weighted reconstructed speech, so as to implement the speech signal compression. In this procedure, the fixed code vector excitation has less impact on the reconstruction of speech signals, which means that the parameters of the fixed code vector can be used for information hiding.

Comparing these parameters, the pitch delay (pitch period) of each frame is encoded with13 bits, and a parity bit is added for checking the pitch delay of the first frame. That means the parameter, pitch delay, has an important impact on reconstructing the speech signals. The pitch delay error, caused by noise interference, is not allowed.

From this analysis, we find that different parameters in G.729 [51] have various capabilities of noisy resistance, which reveals the capacity of carrying secret information. Experiments are carried out to estimate the CNT for parameters of the G.729 codec further by using the method of objective quality evaluation. Twenty 10-second speech sample files were employed as cover objects. In our experiments, the CNT of the parameter of fixed codebook index, fixed codebook sign, gains vector quantization (stage 2), and second stage vector of the LSP quantizer were evaluated. Two types of experiments were carried out. The speech parameters encoded by G.729 are inversed bit by bit or in a row in order to evaluate the impact that every bit of these parameters has.

The $DSNR$ value is used as an objective evaluation standard for speech quality [41,122]. The $DSNR$ is defined as the difference in SNR between the original and the stego speech, given by

$$DSNR = |SNR_b - SNR_a|, \tag{8.1}$$

where SNR_b and SNR_a are the SNR of the carrier speech, and the stego speech respectively. The parameters with a $DSNR$ value of less than 0.5 dB have a small impact on speech quality, and can be used for embedding information [83]. The experimental results are shown in Figures 8.4 and 8.5.

The parameters of fixed codebook index and fixed codebook sign have a small impact on *DSNR*. In Figure 8.4 it can be seen that every bit of parameter L3 has little influence on *DSNR*. From the curve trend of LSP in Figure 8.5, the curve becomes smooth after 5 bits of L3 are accumulated inverted, which confirms the earlier result. We can conclude that the fixed codebook index parameter and the second stage

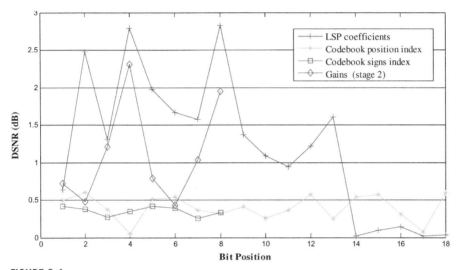

FIGURE 8.4

DSNR value of single-bit inversion.

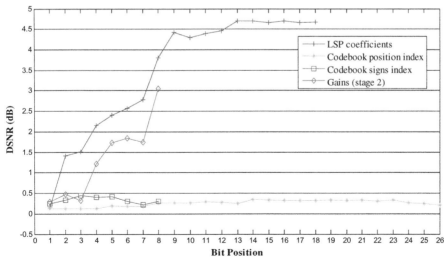

FIGURE 8.5

DSNR of accumulation inversion.

higher vector of the LSP quantizer can be chosen to embed information, while the fixed codebook sign can be used for hiding if the demand of imperceptibility is low but the requirement of embedding rate is high.

Based on this analysis of CNT, we can conclude that there is seldom redundancy in the G.729 encoded speech due to digital compressing. But errors in some parameters may have an imprudent impact on speech quality, as well as the network packet loss.

8.3.2 EMBEDDING APPROACH BASED ON MATRIX CODING

The n-order augmented identity matrix [143] is composed of an n-order identity matrix and an all-one column vector. This kind of matrix may be called an n-order binary augmented identity matrix. The $2L$-order augmented identity matrix can be called a generator matrix. Let

$$
\mathbf{A} = \begin{bmatrix} 1 & 0 & 0 & \dots & 0 & 0 & 1 \\ 0 & 1 & 0 & \dots & 0 & 0 & 1 \\ \vdots & \vdots & \vdots & \dots & \vdots & \vdots & \vdots \\ 0 & 0 & 0 & \dots & 0 & 1 & 1 \end{bmatrix}_{2L \times (2L+1)}. \tag{8.2}
$$

It is assumed that one discretional column vector can be x, which is a $2L$-order binary vector.

If 1 in x is j, $1 \le j \le L$, the row position 1 is used to obtain the vectors in L column corresponding to the 1 value vector in the first $2L$ columns of $\mathbf{A}$. The number of obtained vectors is less than L. Then x can be generated by these column vectors through modulo-two operation.

If 1 in x is j, $L \le j \le 2L$, then the number 0 should be k, $0 \le k \le L - 1$. Thus, the row position 0 is used to obtain the vectors in $L - 1$ column corresponding to the 0 value vector in the first $2L$ of $\mathbf{A}$. The number of vectors obtained is less than $L - 1$. By adding the all-one vector in the last column, x can be generated by these column vectors through modulo-two operation.

From this analysis, for any $2L$-order binary column vector x, the vectors of L column (the total number is less than L) corresponding to a $2L$-order augmented identity matrix will always be x. x is computed by modulo-two operation with these column vectors. That is,

$$
a_{i1}, a_{i2}, \dots, a_{ij}, \quad 1 \le j \le L. \tag{8.3}
$$

$a_{i1}, a_{i2}, \dots, a_{ij}$ are binary column vectors that satisfy Eq. (8.3).

This method can embed $2L$ bits of secret information into $2L + 1$ bits of host information by modifying no more than L bits of host information.

Let $m = (m_1, m_2, \dots, m_n)^{\mathrm{T}}$ be the cover vector that can be modified to embed secret information and $s = (s_1, s_2, \dots, s_{2L})^{\mathrm{T}}$ be the secret information to be embedded. Defining $\mathbf{G}$ as the $2L$-order generator matrix, two conditions are obtained [143,144], $n = 2L + 1$ and $n > 2L + 1$, as explained next.

$n = 2L + 1$

1. Embedding process. Calculate the intermediate vector

$$
b = \mathbf{G}m = m_1 g_1 \oplus m_2 g_2 \oplus \cdots \oplus m_{2L+1} g_{2L+1}, \tag{8.4}
$$

where $g_i(i = 1,\cdots,2L+1)$ represents the i th column of the generator matrix. Then use the secret information s to compute

$$x = b \oplus s. \tag{8.5}$$

If $x = 0$, then $b = s$, and no modification needs to be done to cover vector m. If not, it means that $x \neq 0$, $g_{i1}, g_{i2}, \cdots, g_{ij}$ satisfies Eq. (8.6) according to the aforesaid coding principle:

$$x = g_{i1} \oplus g_{i2} \oplus \cdots \oplus g_{ij} \tag{8.6}$$

and j $(1 \leq j \leq L)$ column vectors are found.

Then the stego vector m' with embedded information can be obtained by negating the j bits in the cover vector m, whose positions correspond to $g_{i1}, g_{i2}, \cdots, g_{ij}$.

Thus, by modifying at most L bits of $2L + 1$ bits to cover the information vector, $2L$ bits of secret information can be embedded.

2. Extraction process. The stego vector m' with embedded information can be expressed as

$$m' = (m_1, \cdots, m_{i1} \oplus 1, \cdots, m_{i2} \oplus 1, \cdots, m_{ij} \oplus 1, \cdots, m_{2L+1})^{\mathrm{T}} \tag{8.7}$$

and

$$Gm' = m_1 g_1 \oplus \cdots \oplus (m_{i1} \oplus 1)g_{i1} \oplus \cdots \oplus (m_{i2} \oplus 1)g_{i2} \cdots \oplus (m_{ij} \oplus 1)g_{ij} \oplus \cdots \oplus m_{2L+1}g_{2L+1}$$
$$= [m_1 g_1 \oplus \cdots \oplus m_{i1}g_{i1} \oplus \cdots \oplus m_{i2}g_{i2} \oplus \cdots \oplus m_{ij}g_{ij} \oplus \cdots \oplus m_{2L+1}g_{2L+1}] \oplus [g_{i1} \oplus g_{i2} \oplus \cdots \oplus g_{ij}].$$

Combining Eqs. (8.4), (8.6), and (8.7), it can be inferred that

$$Gm' = b \oplus x = b \oplus b \oplus s = s. \tag{8.8}$$

It is obvious that the original secret information s can be easily obtained by executing a modulo 2 operation to G and m'.

Based on the preceding discussion, only the j column vectors in generator matrix G are necessary to satisfy Eq. (8.6), and the sequence position of column vectors in G can be discretional, thus the order of G is chosen as $(2L + 1)!$.

$n > 2L + 1$

From the condition of $n = 2L + 1$ we learn that the G satisfying $n > 2L + 1$ is a $2L \times n$-dimension matrix. In order to make the binary vector x meet Eq. (8.6), each column vector of A (see Eq. (8.2)) must be used at least once. The rest of the column vectors can be formed by any column vector of A, and the selections of G are $(2L+1)! \times C_n^{2L+1} \times (2L+1)^{n-2L-1}$. If n is large enough, the number must be great. Therefore G can be used as the embedding key. Embedding may be varied in accordance with different generator matrices G while the extraction process remains the same [143,144].

Based on this analysis, we can see that there are a large number of selections of generator matrix G, which is used both in the embedding process and the extraction process. G can be regarded as an embedding key. To improve the security of a covert communication system, a different G may be taken into consideration for odd frames and even frames.

In addition, interleaving technology is a kind of channel coding technique. It is usually used to fight against cluster error bits caused by multipath and fading in mobile communication channels. Interleaving decreases the correlation between each codeword. Therefore clusters of errors in the transmission channel can be transformed into random errors. This means that errors are dispersed, improving the reliability of the system. To improve the antiattack ability and reduce bit error rate in extraction, interleaving technology may be a candidate for enhancing the robustness of the embedding approach.

8.3.3 EMBEDDING PROCEDURE

Based on the analysis for CNT and experimental results, L3 in LSP, and partial C1 and C2 in fixed codebook [142] index bits are chosen to form the cover bits. The explicit cover bits selection scheme is illustrated in Figure 8.6.

An embedding flow chart and explicit operation are presented respectively as follows.

The basic procedure of the embedding scheme is shown in Figure 8.7. To improve the security of secret speech, MELP-encoded secret speech is first encrypted by a pseudorandom sequence. Then embedding positions and the cover vectors are calculated by using the generator matrix **G**. Finally, the processed bit stream in the public communication channel is transmitted. The receiver conducts the inverse process of the sender. The generator matrix **G** is used for extracting secret bits from stego speech. Then the extracted bits are decrypted to restore the original secret speech [142–144].

Considering that the G.729 codec outputs 80 bits as a frame (10 ms) and MELP codec outputs 54 bits as a speech frame (22.5 ms), 9 frames of G.729 encoded speech are set as one embedding unit to embed four frames in MELP-encoded secret speech. Hence two speech bit streams can be synchronized in a time scale; for example, 90 ms of secret speech is embedded into 90 ms of carrier speech. The interleaving and embedding procedures are shown in Figure 8.8 [142–144].

The required embedding rate is relatively high. The embedding process is completed by using the matrix coding method (introduced earlier). In the proposed embedding scheme, the number of cover bits in each frame is $n = 25$ and the number

L0	L1	L1	L1	L1	L1	L1	L1
L2	L2	L2	L2	L2	L3	L3	L3
L3	L3	P1	P1	P1	P1	P1	P1
P1	P1	P0	C1	C1	C1	C1	C1
C1	C1	C1	C1	C1	C1	C1	C1
S1	S1	S1	S1	GA1	GA1	GA1	GB1
GB1	GB1	GB1	P2	P2	P2	P2	P2
C2	C2	C2	C2	C2	C2	C2	C2
C2	C2	C2	C2	C2	S2	S2	S2
S2	GA2	GA2	GA2	GB2	GB2	GB2	GB2

☐ Normal parameter position   ▨ Selected parameter to carry secret data

FIGURE 8.6

Cover bits selection scheme.

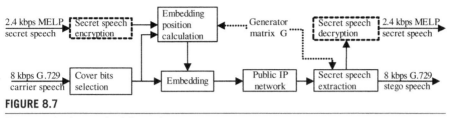

FIGURE 8.7

Framework of an embedding scheme.

of secret bits to be embedded is $2L = 24$, which satisfies the condition of $n = 2L + 1$. Thus with no more than 12 bits of host information modified, 24 bits of secret data are embedded in 25 bits of host information. The embedding rate reaches $R = 24 \times 1000/10 = 2.4$ kbps, which meets the requirement of embedding 2.4 kbps MELP encoded speech into 8 kbps G.729 encoded speech [51].

The following gives the detailed steps of embedding and extraction.

8.3.3.1 Embedding process

Use the 2.4 kbps MELP speech codec to encode the secret speech and encrypt the coded bit stream by m sequence. Regroup the encrypted bits to construct matrix **S**. In accordance with Figure 8.8, perform the preprocess by interleaving, and pick column vectors of each frame to build the secret vector, which is denoted as $s = (s_1, s_1, \ldots, s_{24})$.

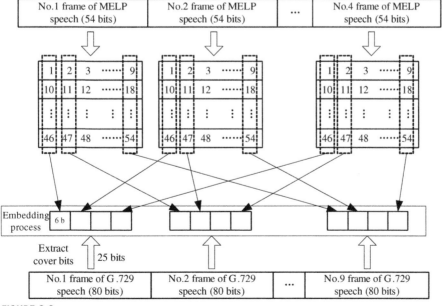

FIGURE 8.8

Detailed operations of interleaving and embedding.

For each frame of G.729 encoded carrier speech, 25 bits are selected to build a cover vector $m = (m_1, m_2, \ldots, m_{25})$ according to the detailed operation shown in Figure 8.8 and cover bits selection scheme in Figure 8.6. The generator matrix $\mathbf{G}$ consists of a 24-order augmented identity matrix and an all-one column vector. Calculate $b = \mathbf{G}m$ according to Eq. (8.4), and then compute $x = b \oplus s$ according to Eq. (8.5). If x is an all-zero vector, no modification is done on the cover vector. If not, modify m in accordance with Eqs. (8.6) and (8.7) and obtain the modified vector m', which carries the secret bits of s.

Substitute the corresponding parameter bits in the original G.729 coded bit stream according to m' and transmit the processed bit stream in the public channel.

8.3.3.2 Extraction process

According to Figure 8.8, extract the preselected parameter bits in each frame of the stego bit stream to reconstruct the modified vector m'. Then calculate $\mathbf{G}m'$ to obtain the secret vector s.

Decrypt s by using the m sequence. The m sequence must be the same as that used in the embedding process. Decode the decrypted bit stream with the MELP decoder to reconstruct the original secret speech.

8.3.4 EXPERIMENTAL RESULTS AND ANALYSIS

Simulation experiments are performed on a Windows XP computer. Both the original carrier speech and the secret speech are sampled with an 8000 Hz sampling rate. Twenty different speech samples are involved in the experiments, quantized with 16 bits and encoded by linear PCM [66]. Both the objective measurements and subjective listening tests are conducted on mixed speech to evaluate the proposed method's embedding effect.

8.3.4.1 Objective Quality

The *DSNR* value is used as an objective evaluation standard for speech quality. Figure 8.9 shows the results of the experiment on the sampled 20 speech files, with the horizontal axis representing the speech sample index, and the vertical axis representing the *DSNR* values of the difference in *SNR* between the original speech and the stego speech. The results indicate that the *DSNR* values between the original speech and the stego speech are usually small to ensure a certain embedding capacity. The average *DSNR* value of 20 speech decreases around 1.2 dB. Figure 8.9 shows the *DSNR* result.

8.3.4.2 Subjective Quality

The ITU P.862 recommendation is adopted to measure the subjective quality of the stego speech. The recommendation describes an objective method for predicting the subjective quality of the speech codec. It uses the perceptual evaluation speech quality (PESQ) value to assess the subjective quality of the stego speech. According to the ITU P.862 standard, the PESQ-MOS value of the original speech is about 3.49 [147]. The average PESQ-MOS value of the stego speech is estimated to be 3.11. The difference in PESQ-MOS value between the original speech and stego speech is too minor, so the imperceptibility of the steganography algorithm is acceptable. Figure 8.10 shows the PESQ-MOS result.

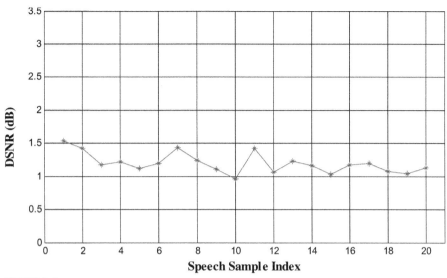

FIGURE 8.9

DSNR for carrier speech and stego speech.

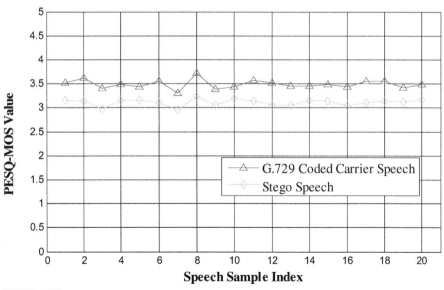

FIGURE 8.10

PESQ-MOS values of carrier speech and stego speech.

For evaluating the imperceptibility of the stego speech further, the comparison of original speech with stego speech is completed in the time domain and frequency domain. One of the 20 speech files is chosen as an experimental object for information embedding and extraction. The comparison results are shown in Figures 8.11 and 8.12.

The comparison result is shown in Figure 8.11. For the sake of the parameter coding scheme, the waveform of G.729 coded speech cannot be exactly the same as PCM sampled speech, but intelligibility and naturalness are preserved. To some extent, the spectrogram may reflect voice characteristics, such as the formant feature of vocal cords, pitch frequencies, and flags of voiced and unvoiced segments. It is a fact that the waveform changes a little when embedding occurs. Stego speech changes a

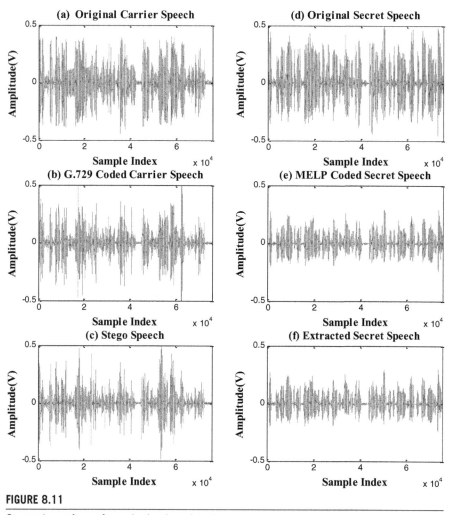

FIGURE 8.11

Comparison of waveforms in the time domain.

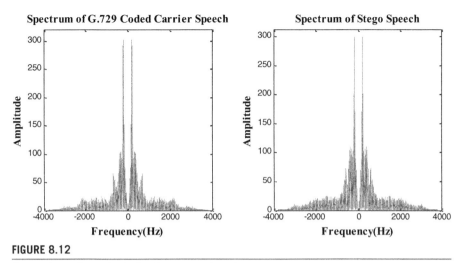

FIGURE 8.12

Comparison of spectra in the frequency domain.

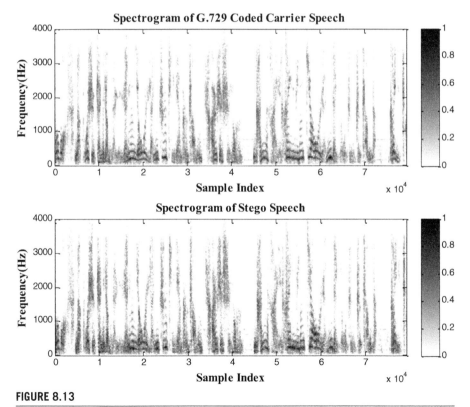

FIGURE 8.13

Spectrograms of carrier speech and stego speech.

little in the frequency domain too. From the viewpoint of the spectrogram, stego speech maintains the basic features of the original carrier speech, and the harmonic wave in the frequency direction is still clear (see Figure 8.13). In addition, the time-varying process can also be seen in the spectrogram. Based on the previous results analysis, this embedding method has no sensible impact on carrier speech, and it also has good performance in imperceptibility. Comparing the waveforms of MELP-coded secret speech with extracted secret speech (see Figure 8.11), results show that they are almost exactly the same, which means that the bit error rate of extraction approaches 0.

8.4 SUMMARY

An algorithm for embedding secret speech into public speech is proposed for the purpose of secure communication over VoIP in this chapter. This algorithm uses bits of the second stage higher vector of the LSP quantizer to carry secret speech by performing CNT on each parameter bit of G.729. A 2.4 kbps MELP-coded speech is embedded into 8 kbps G.729-coded speech by combining matrix coding and the interleaving technique. Experiments on embedding capacity, rate, and time complexity have been conducted, and the results show that a high data embedding rate up to 2.4 kbps with better imperceptibility are achieved. Comparing the proposed algorithm with G.729-based embedding methods, the results prove that the proposed algorithm achieves good embedding performance with low complexity, which may make real-time embedding/extracting possible. In future research, the steganalysis of the proposed algorithm will be carried out, and the implementation of practical use in the VoIP system will also be studied.

Design of Real-Time Speech Secure Communication over PSTN

Because digital media are gaining wider popularity, their security-related issues are more concerning. Nowadays, information hiding techniques are widely used in communication systems. A new steganography scheme of realizing Speech Information Hiding Telephony (SITH) for hiding one piece of speech information data into another is proposed based on an information hiding security technique. In this technique a host speech as a carrier plays a role in two ways: first, it provides a common communication channel for common information; second, it sets up a covertures channel for steganography information (secret message). Speech information hiding encryption communication is a system based on secure steganography techniques. SITH is different from an encryption telephone; it is designed for achieving privacy when speech is transmitted and communicated over PSTN.

9.1 SECURE COMMUNICATION PLAN

In this section, the scheme of real-time speech secure communication over PSTN is introduced. The design of SIHT using speech information hiding technology is presented.

9.1.1 INTRODUCTION

Nowadays security in data transmission is especially important for commercial telecommunications. Hence a secure communication system is essential to people's lives. A secure communication system should have the following characteristics [50,81,86]:

- Good performance in speech hiding. This system adopts the proposed ABS embedding and extraction algorithm. It hides low bit-rate secret speech into high rate carrier speech by bit, rather than hide secret speech at the beginning or end of the carrier. This hiding mechanism performs very well under the situation of format conversion or when being attacked. The carrier speech is coherent and meaningful, and the secret speech is encrypted. In public channels, composite speech sounds like the carrier, and thus it is not likely to attract eavesdroppers' attention. Different carrier speeches can be recorded in advance to form a candidate carrier set.

Information Hiding in Speech Signals for Secure Communication. DOI: 10.1016/B978-0-12-801328-1.00009-4

- Voice quality guaranteed. The extracted and decoded secret speech does not degrade in quality, and distortion is very small. The continuity and naturalness of speech must be ensured (human ears can hardly sense the difference between original and extracted speech).
- Enhanced security in authentication and encryption. The system employs a challenge-response authentication method. Hence, the sender and receiver may confirm the correspondent's legal identity to establish a reliable and secure communication channel. Meanwhile, the hidden secret speech has also been encrypted by chaotic sequence to avoid being heard by eavesdroppers.
- Good performance in real time. Because the system adopts a parallel processing structure constituted by a DSP array, the speech encoding, decoding, encryption, decryption, hiding, and extraction process speed can meet the requirements of real time.

9.1.2 REQUIREMENTS ANALYSIS

The main functions of the speech secure communication system are to implement safe and reliable voice/speech secure communication. It works on a two-way full-duplex mode of communication.

- **Technical Indicator.** The technical indexes are composed of speed, delay, and BER (Bit Error Rate).
  - Speed—In a speech information hiding secure communication system, the speech signal is sampled at a rate of 8 kbps, and one transmitted frame contains 160 samples (20 ms, 4 subframes, 5 ms each subframe). To ensure the real time of speech communication, the required speed of data communication should be more than 12.8 kbps [41,122].
  - Delay—Requirement for the delay is only a theoretical estimation, and calculation includes computational processing delay, transmission delay between DTE (Data Terminal Equipment), DCE (Data Communications Equipment), and so on.
- Taking the reference data into consideration, it is assumed that:
  - Whenever data reaches the computer, there is a processing time slice to deal with data.
  - Computing processing delay (including the detention between DTE and DCE and the time that the data arrives at application level process) takes 10 ms.
  - The application level processing time is 100 ms (typically in the FILE TRANSAFE type of application, it needs a steady stream of data).
- With regard to the transfer rate of 12.8 kbps, the maximum allowable delay is less than 1 s.

  - BER—For real-time voice/speech and data transmission, BER should be less than 10^{-7}, which is the same as the requirement for normal data communication.

- **Functional Requirements.** The functional requirements consist of security, reliability, and secrecy.
  - Security—The main feature of a speech information hiding secure communication system is high security. There are a number of PSTN-based voice communication lines [148], voice lines, or Internet modem lines or fax lines, and many of them are occupied simultaneously. Ideally, speech information hiding technology makes the secure communication line completely the same as the ordinary lines, to ensure the secrecy of communication [149]. In addition, the system's security is contributed mainly by the design features of the hiding algorithm.
  - Reliability—The system adopts coordinate communication mode. Measures to ensure the reliability are:
    - Frame mode. Using frame mode, the header is used to locate the beginning of a frame. More importantly, by examining the header, if it is correct, the analysis continues.
    - CRC check. Check the CRC of the whole data frame to ensure the accuracy of the header. The CRC checksum is 16 bits, and thus the possibility of error is minimal.
  - Secrecy—The data secrecy of this system mainly relies on the encryption strength of the chaotic sequence and its own security.

9.2 DESIGN AND REALIZATION OF A SECURE COMMUNICATION SYSTEM BASED ON PC

Based on the research of speech information hiding, the MELP 2.4 kbps coding scheme is selected as the secret speech coding method, while the carrier speech is coded by the GSM scheme. According to the information hiding algorithm based on GSM, the secret speech bits are embedded into carrier bits in real time, to build a speech security communication test system that adopts PC and MODEM. Eavesdroppers can only sense the modulated digital signal but cannot determine whether the signal contains secret information, even if they demodulate the received bits. This system has achieved the goal of confusing eavesdroppers with strong concealment and security [25,138].

This system is developed in the VC^{++} environment, and the application platform is Windows or NT. Crucial algorithms are programmed by standard C, to facilitate porting to a DSP hardware platform.

9.2.1 FRAMEWORK FOR DESIGN

A data transmission system based on PSTN includes MODEM, ISDN, ADSL, and so on. The MODEM method is selected for data transmission due to the simplicity

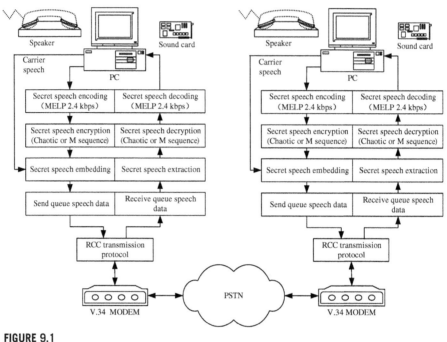

FIGURE 9.1

Framework of a secure communication system based on PC.

of the device, purchase, installation, and usage. The overall structure of the system is shown in Figure 9.1 [86,87,90].

Each part of the communication system has a V.34 MODEM operating at the rate of 56 kbps, which is connected to PSTN through an analog telephone line. In practical applications, MODEM employs the RCC (Reliable Comm Communication) protocol to transmit data, with a maximum transmission rate of 28 kbps.

An experimental system is implemented on the Windows platform. The PC is connected with MODEM through serial ports, and the interrupt-driven serial communication driver COMM.DRV is provided by Windows. The communication program does not operate on serial ports directly, but an indirect action is taken through the API. In Windows, serial communication uses *event notification* mode to support data transfer in blocks. At the beginning of communication, the operating system opens up a user-defined input/output buffer. A low-level hardware event occurs when it receives a character, and hence the serial driver gets control access immediately, which sends the character to input buffers. Control access returns to running applications. In the receiver, if the input buffer is full, the driver program utilizes a present-defined stream control mechanism to inform the sender to stop transferring. The sender adopts a method similar to that employed by the receiver to avoid buffer overflow.

The experimental system operates in full-duplex mode, and the equipment on both communication parts are ordinary telephones. The work process of the experimental system is as follows [90]:

1. Initialize MODEM and dial the other side's number to build a communication link.
2. Collect secret speech by using a sound card, then encode and encrypt the collected secret speech by using the selected low-rate speech coding and encryption scheme.
3. Embed secret speech into the public carrier speech by use of the proposed ABS embedding algorithm to generate composite speech bit-stream.
4. Transmit composite speech from sender to receiver through MODEM, and the other side's MODEM can receive and demodulate composite speech bit-stream.
5. Extract secret speech by use of the proposed ABS extraction algorithm.
6. Decrypt the extracted signal to obtain plain text, and then the decrypted signal is decoded by using a low-rate decoding scheme to produce recovered secret speech, which is played by the sound card.

The quality of reconstructed speech has degraded a little compared with that of original speech. Accompanied by little noise, the third party could completely regard it as normal communication without any reference. It is difficult for the third party to sense the embedded secret data. Hence the purpose of information hiding is achieved.

9.2.2 CODING SCHEME SELECTION

In the experimental system, selection of the GSM (RPE-LTP) coding scheme is used as a carrier coding method mainly for these reasons [137,138]:

- The GSM coding scheme has a relatively low bit rate (13 kbps), and the reconstruction of the synthetic speech retains good quality.
- Some parameters have relatively strong robustness. Therefore, a small change to these parameters does not greatly influence the quality of reconstructed speech.

The MELP 2.4 kbps coding scheme is used as the secret speech coding method. According to the proposed ABS model and the characteristics of GSM coding, the technique of the information embedding and extraction algorithm based on the GSM coding scheme is presented. The embedding capacity test shows that the constituted algorithm can reach a maximum capacity of 2.6 kbps, and the quality of reconstructed speech is good.

9.2.3 MULTIMEDIA PROGRAMMING

An experimental system for PC-based speech security communication is on a Windows platform. Voice processing adopts the API function of Windows, sound recording utilizes MCI functions, and sound playing uses low-level audio functions.

MCI functions can be used to control access to the recording/playing of WAV files, whereas the low-level audio functions may complete these two major tasks [150,151]:

- Controlling access to GSM and MELP files as well as the recording/playing of PCM files.
- Ensuring that the sender and receiver can process real-time transmission.

The MMIO function can be used to create and store WAV files, and it can also be used to open a WAV file.

The structure of multimedia systems is divided into three levels:

- The bottom layer is the control and driver of the sound device. At this level, the audio function and MCI function are used to open up the sound device and set parameters.
- The middle layer is data processing, including settings of the buffer and the access of data.
- The top layer is applications; for example, the encoding/decoding and encryption/decryption processes.

9.2.3.1 Implementation Levels
The graph of implementation levels is shown in Figure 9.2 [150,151].

9.2.3.2 Program Flow
There are two branches of the program processes: file processing and data buffer processing. File processing is accomplished by the MCI function and data buffer processing is realized by a low-level audio function. The flow chart is shown in Figure 9.3 [150,151].

It should be noted that, in Figure 9.3, playback for WAV, PCM, GSM, and MELP formats is treated in different ways. WAV is a kind of RIFF block standardized multimedia file format. Hence the voice parameters (i.e., sampling rate, sampling accuracy, etc.) are included in the header. If you use MCI for playing, those parameters can be identified from the information in the header, and programming is quite

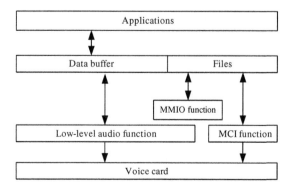

FIGURE 9.2

The schematic diagram of implementation levels.

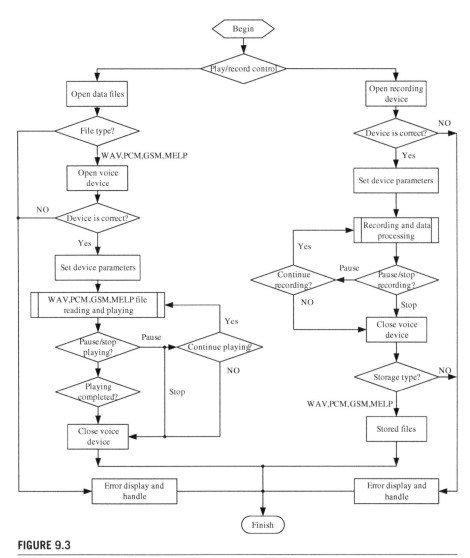

FIGURE 9.3

The file processing flow chart.

simple. For the other three format files, voice parameters must be specified. As a general requirement, carrier speech should be conveyed with a sampling rate of 8 kHz, sampling accuracy of 8 bits, and mono track [41,122].

The flow chart of the data sending buffer is shown in Figure 9.4 [150,151].

PCM-coded files can be played directly while GSM- or MELP-coded files should first be decoded into PCM and then played. To improve decoding speed, multibuffer blocks may be utilized, and thus decoding and playing can be executed simultaneously [41,122].

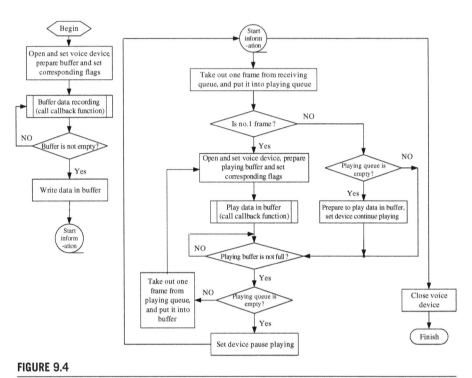

FIGURE 9.4

The data sending buffer flow chart.

So far, neither the details of voice device settings nor the buffer settings have been given. Attention should be paid to the processing levels. The top is the sending/receiving queue, the bottom is the buffer queue of the voice device, and the middle is the recording/playing queue, which is used for temporal data storage. The middle level plays an essential role in data transform and data protection. Because the recording and playing occur at the same time, multithread is needed. Because of real-time sending/receiving, the data queue cannot be accessed simultaneously, otherwise unpredictable damage may happen to data. The adoption of a middle level and the establishment of flags are good solutions to this problem.

9.2.4 SYSTEM REALIZATION

An experimental system for PC-based speech secure communication has the general functions to meet the requirements of ordinary users. In practical systems, corresponding functions can be added based on users' requirements. The Graphical User Interface (GUI) is shown in Figure 9.5.

The whole system is developed in a VC++ environment, and the key algorithms are programmed by standard C. The main modules are RCC communication, speech collecting/playing, speech encoding/decoding, data embedding/extraction module and data encryption/decryption.

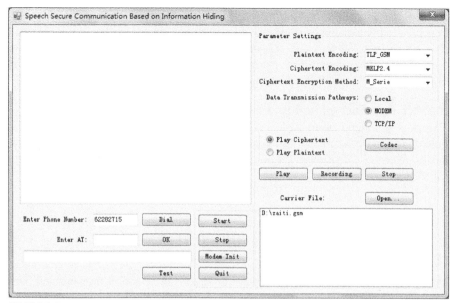

FIGURE 9.5

The GUI of a PC-based speech secure communication experimental system.

9.2.4.1 RCC Communication Module

The RCC communication module is mainly responsible for data packing/sending and receiving/unpacking processes. Defined classes, including *CcommReadThread*, *CcommWriteThread*, *CMyComm*, and *CRCCComm*, can accomplish complete functions. *CcommReadThread* monitors serial ports for the receiving buffer, and reads all the data once the data appear in the buffer. *CcommWriteThread* monitors serial ports for the sending buffer, and writes data once the buffer is empty. *CMyComm* inherits a subclass from *CRCCComm*, and it is responsible for packing/sending and receiving/unpacking processes. Member object *m_MyAudio* conveys speech collecting and playing. *m_MyMel* and *m_MyGsm* carries on MELP and LTP encoding/decoding processes respectively [152].

9.2.4.2 Speech Collecting/Playing Module

This module mainly completes speech collecting and playback at the same time, including classes of *Caudio*, *CsoundOut*, and *CsoundIn*. *CsoundOut* is used for playback, and *CsoundIn* is used for collecting speech data. Sampling rate and accuracy can be set manually [150,151].

9.2.4.3 Speech encoding/decoding module

GSM coding, G.728, and MELP coding are accomplished in this module, which consists of *CGsm* and *CMelp*.

9.2.4.4 Data Embedding/Extraction Module

Functions sub_blockembed (unsigned char *bitblock, unsigned char *tmp, unsigned char *y) and sub_unblockembed (unsigned char *tmp, unsigned char *y) constitute the module, whose major function is embedding secret speech into carrier speech and extracting embedded speech from composite speech.

9.2.4.5 Data Encryption/Decryption Module

This part is independent; hence the encryption scheme can be selected as required. The majority of the functions are *ChaosSequenceReceive (int BitNum)* or *SeriesMReceive (int BitNum), ChaosSequenceSend (int BitNum)* or *SeriesMSend (int BitNum)*.

9.3 SPEECH INFORMATION HIDING TELEPHONY (SIHT) BASED ON PSTN

Telephony, as a communication device for speech information transmission over PSTN [90,148], is widely used over the world. Since PSTN is widely applied in the areas of national defense, economy, civilization, and research, more and more attention is being paid to the automatic protection of communication and safety transmission of information with the requirement of high quality and multifunction of PSTN [86,87,90]. It is obviously important to ensure secret communication and validly prevent the attack of illegal eavesdroppers. Here, a new steganography scheme based on information hiding techniques is proposed for hiding one piece of speech information in another, and an SIHT system is designed.

9.3.1 INTRODUCTION

SIHT is designed by using one fixed point DSP TMS320C54x, three floating DSPs TMS320C31, and a single-chip microcontroller based on the embedding system method. All DSPs work in parallel in processing speech encoding, decoding, encryption, decryption, hiding, extraction, bus arbitrary control, and so on. SIHT connects to PSTN like a normal telephone, and it has two speech inputs: public speech and secret speech. Secret speech information is first encrypted, and then hidden (embedded) in public speech information with proprietary algorithms individually. SIHT has a transparent operating mode that does not alter the signal and a secure mode accessed upon request of the speaker. SIHT encrypts the secret speech and embeds it in public speech with digital techniques. At the transmission branch, public speech is sampled and coded with an ADPCM scheme at 32 kbps. Secret speech is sampled and coded with a CELP scheme at 2.4 kbps. Embedded speech is interfaced to the line with E.Modem EM5600. Public speech used as a carrier is selected carefully. High speech quality is guaranteed by the coding scheme, especially at the highest transmission rate [86,87].

The keys for encryption are established at the beginning of each transmission. For security reasons, the keys used for ciphering secret speech and the public speech used as the host data set must be changed with each new call, and forced to carry out a user transparent handshake at the beginning of the session. Cryptoanalysis techniques are being used to test the robustness of the encryption and detection techniques adopted to verify the capacity and robustness of the information hiding with different algorithms individually.

9.3.2 DESCRIPTION OF THE SIHT

Instead of completing the telephone set, SIHT is conceived as a black box connecting the user's set and the line, meeting all the requirements for the system plugged to the PSTN (Figure 9.6) [86,87,90,148]. When user A and user B are communicating with each other by SIHT, they select the transparent mode, and the conversation can be heard by eavesdropper C and eavesdropper D. If they switch the operating mode to secure mode, the secret speech information can be heard by users A and B. Meanwhile, the eavesdroppers C and D can only hear the public speech information, which is irrelevant to what the user A or B is talking about. Whether user A or B is talking about secret speech information or not, eavesdroppers C and D can hear only continuous meaningless speech information. This is different from traditionally encrypted telephony, by which eavesdroppers can hear only meaningless noise. So communications can be made by SIHT without easily causing suspicion of C and D. SIHT may relax their vigilance, and can make the communications encrypted and secure.

SIHT is a compact system, portable and user friendly. Sets of algorithms and a specific architecture have been developed. SIHT has the following features [86,87,90].

9.3.2.1 Robustness of Information Hiding

SIHT adopts a classical algorithm with LSB (Least Significant Bit) substitution for speech information hiding. Given a secret speech m_i, the embedding process consists of choosing a public speech as set $\{j_1,\ldots,j_{l(m)}\}$ (i.e., host data set or carrier), and

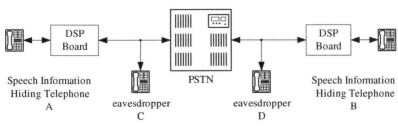

FIGURE 9.6

SIHT connection.

performing the substitution operation $c_{j_i} \Leftrightarrow m_i$ on them, which exchanges the LSB of c_{j_i} by m_i (m_i can either be 1 or 0). We could also imagine a substitution operation, which changes more than one bit of the cover, for instance, by storing two message bits in the two least significant bits of one cover-element. In the extraction process, the LSB of the selected cover-elements is extracted and lined up to reconstruct secret messages [60,153].

In SIHT, an approach based on Discrete Wavelet Transform (DWT) is designed for hiding (embedding) the secret speech data with a CELP scheme at 2.4 kbps in the public speech data and with an ADPCM scheme at 32 kbps bit-to-bit. It does not show the data at the beginning or ending of public speech.

9.3.2.2 Steganography

The basic substitution systems employed in SIHT try to encode secret speech information by substituting insignificant parts of the cover with secret message bits. The receiver can extract the information if he or she has the knowledge of the positions where secret information has been embedded. Since only minor modifications are made in the embedding process, the quality of public speech, in which a secret speech is embedded, would not be degraded. Every time public speech is transmitted over PSTN, eavesdroppers can hear and understand the public speech clearly [90]. But eavesdroppers cannot extract secret speech and cannot clearly understand it, even if they have the awareness of the presence of secret speech. The sender can make sure that a passive attacker will not notice the communication through a covert channel.

9.3.2.3 Secure Encryption

From a security perspective, both public key and private key algorithms are used in SIHT.

Because SIHT includes a data E.Modem type of EM5600t, data can be interchanged between the equipment involved in communications. When one of the users decides to begin a secure communication, the corresponding equipment sends to the other its public key number. Once this number has been received, the other equipment generates a random number (the session key), ciphers it with the public key, and sends it backs, so the first equipment is able to decipher the session key. This key is not stored and a new key is generated for each call.

SIHT also includes an algorithm for generating public keys and therefore these keys can be replaced periodically in order to enhance the global security of the system.

9.3.2.4 Speech Quality

In SIHT, public speech signals are coded with an ADPCM scheme at 32 kbps, and this speech with embedded secret speech is transferred over PSTN. The quality of public speech degrades slightly at 32 kbps with secret speech embedded in it, but preserves perfect intelligibility and speaker characteristics. This is measured by a series of informal tests on a board and heterogeneous group of end users. The analytic speech frame has a length of 20 ms with four subframes of 5 ms.

9.3.2.5 Real Time

Parallel processing is implemented in SIHT. Speech coding, decoding, encryption, decryption, embedding, and extraction processing are handled at the same time in different ways by three DSPs individually. SIHT is low-delayed and synchronous [154].

9.3.3 SPEECH INFORMATION HIDING SCHEME

Based on analysis of speech signals in SIHT [148], a fast and secure key has been designed to encrypt secret speech information. The secret speech information is then hidden in public speech information using an information hiding algorithm, without obvious degradation to public speech quality. Public speech is continuous, meaningful, and irrelevant to secret speech. Public speech, with secret speech embedded in it, is transferred from a sender to a receiver. At the receiver end, an algorithm of secret speech detection and extraction must be applied to obtain the primary secret speech information.

Inside SIHT, a speech information hiding scheme (Figure 9.7) [86,90] has been implemented using the MATLAB software package. Simulation steps are as follows (Train Whistle as public speech, Dropping Egg as secret speech):

1. Two speech signals and a key are the input signals. The key is used to encrypt coded secret speech information, and then the encrypted secret speech information is embedded into public speech information at the end of the speech information hider. Then the combined speech information is transferred over PSTN.

2. Public and secret speech can be selected as output signals. Secret speech has been detected and extracted in a speech extractor after decoding. It is then sent to the speech switch. Public speech bypasses the decoder to the switch too.

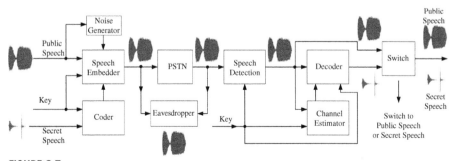

FIGURE 9.7

Speech embedding scheme.

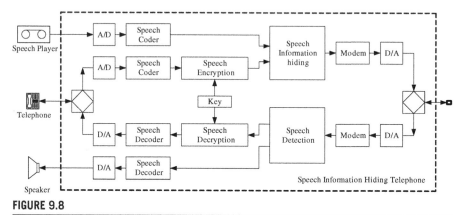

FIGURE 9.8

SIHT module.

9.3.4 **SIHT MODULE**

The main block distribution of SIHT is depicted in Figure 9.8. Briefly, it consists of two blocks: a transmitter and a receiver [86,87,90].

An analog to digital converter has 16-bit resolution with a sample rate of 22.05 kHz. Speech coding must be carried out by sophisticated algorithms in order to satisfy it with a moderate bit rate of 2.4 kbps. There are options for key management. The basic key management process avoids human intervention by means of a public key generation algorithm (RSA) and a session key for the speech encryption algorithm, performed through a private algorithm. For speech transmission, the E.Modem is based on an adapted version of the V. 90 CCITT recommendation. This modem works at full-duplex 33.6 kbps, using a QPSK coded modulation. The modem implementation is entirely digitalized. The following deals with the system components of SIHT in detail [86,87,90].

9.3.4.1 Speech Coder and Speech Decoder

Speech coders used in SIHT are CELP 2.4 kbps type for secret speech and ADPCM 32 kbps type for public speech. Speech quality degrades slightly, but preserves perfect intelligibility and speaker characteristics [41,122].

The analysis speech is updated once a frame. The rest of the coder parameters are long-term predictor, stochastic excitation, and vector gain. They are updated four times per frame by using subframes with a length of 5 ms.

The short-term predictor is a 10-order LPC filter, obtained by the autocorrelation method and hamming windowing. There is no overlap between analysis frames but four sets of interpolated coefficients are obtained to avoid rough transitions in the synthetic speech spectrum. For transmission and interpolation, LPC coefficients are transformed into line spectrum pairs (LSPs). The LSP coefficients are scalar quantifiers. The quantification is different between successive LSPs. The transmission is done in an efficient way by locating transitions over the quantification values [154]. Prior to the transformation, a 15 Hz spectra expansion is applied.

For every subframe, a closed-loop analysis is performed to get the index and the gain of the adaptive excitation. After this in the 2.4 kbps coder, a three-order long-term predictor is calculated, and the improvement in subjective quality of synthetic speech with the three-order predictor is noticeable. The stochastic library contains 512 vectors. It is sparse (80% zero values) and overlapped with a shift of two samples.

All the quantification values are obtained by training an LBG quantizer with 14 minutes of processed speeches, which are collected from 7 female and 7 male speakers [154].

There are sets of bits in every frame for error detection. The protected bits are the most sensitive bits obtained in simulations.

An adaptive algorithm based on the coder parameter is used to suppress the background noise introduced by digital-to-analog conversion in the silence period. It also cancels the echo introduced by the hybrid circuits.

9.3.4.2 Encryption and Decryption

The generation and distribution of session keys are the most difficult in a cryptographic system, although many solutions have been proposed. SIHT uses an automatic method, which is transparent to users and eavesdroppers and based on a public key encryption algorithm, to avoid erroneous manipulations.

Because SIHT includes a data E.Modem, data can be interchanged between the equipment involved in the communication. When one of the users decides to switch to secure mode, the corresponding equipment sends its public key number to the other. Once this number is received, the other equipment generates a random number (the session key), ciphers it with the public key, and sends it back to the first equipment, which is the only one able to decipher the session key. In this way both pieces of equipment have the session key, which is not stored, and a new one is generated for each call [41,42].

SIHT also includes the algorithms for generating the public keys and therefore these keys can be replaced periodically in order to enhance the global security of the system.

In the following explanation, the detailed robust ciphering procedure is introduced.

Once the E.Modems of both sides have established a link, both microcontrollers generate a random number of 32 bits. Then, they send their 32-bit public key to each other. As these are public keys, their transmission does not affect the security of the system. Each device uses the other's key to encrypt the previously generated random number, following a variant of the RSA algorithm (Figure 9.9) [86,87,90]. Even though the calculation is rather exhaustive for an 8-bit processor, the low amount of information makes it feasible in real time. Now the 32-bit random number, ciphered with 32-bit public keys, is shared over the PSTN [148]. According to Shannon' theory [106,107], the fact that both the key and the message have the same length assures perfect secrecy.

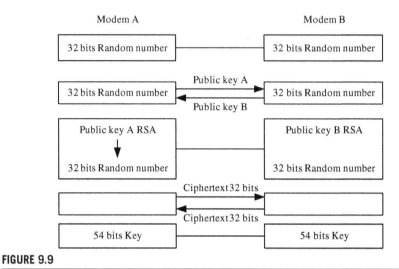

FIGURE 9.9

RSA-like encryption scheme.

Each device decrypts the received message with their private keys. Hence, by now, both microcontrollers know each other's 32-bit randomly generated key. The two 32-bit random numbers are combined to generate a 64-bit key, which is used to encrypt speech (only 54 bits are used for encryption; the other 12 bits are used for verification). All the processes have been performed during the transition to protected mode, following the modem handshaking. Required calculus is heavy, but time delay is negligible since message length is short.

For real-time operation, an RSA-like approach for speech encryption is not feasible to the available controller, so a simplified scheme is proposed (Figure 9.10) [90].

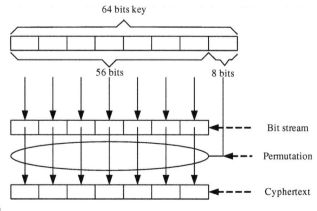

FIGURE 9.10

Permutator-based encryption scheme.

This second procedure is based on permutations and substitutions similar to DES, but with fewer numbers of stages to reduce processing. Specifically, the first 56 bits in the key are XORed with the incoming bytes of information, using a circular buffer to rotate the key. The last 8 bits of the key define the permutation to be applied to every encrypted byte. This method achieves a high degree of security and low computational complexity because the redundancies in the coded signals have been minimized.

9.3.4.3 E.Modem

SIHT utilizes an embedded modem EM5600 developed by Beijing Comsys Technologies Co. Ltd. The ASICs manufactured by Rockwell are adopted in this module. The EM5600 module supports the standard AT command set. In addition, it has functions of dial-up and auto-answer. EM5600 has the following features:

- Speed of transmission: 14.4 kbps, 33.6 kbps, and 56 kbps. Supporting CCITT V.90bis, V.34, and V.90
- Support of the standard AT command set
- Support of dial-up and auto-answer functions
- Dual-lined SIP14 pins that physically interface with target system
- Two types of interfaces with user's systems: 8 bits parallel interface and RS-232 interface.

9.3.4.4 Hider and Extractor

The technique of information hiding is implemented in SIHT, which mainly focuses on two aspects, secret speech information concealing and public speech information hiding. The latter is necessary and important.

Whereas cryptographic techniques try to conceal the contents of information, information hiding (steganography) goes a bit further: it tries to conceal not only the contents of information, but also its very existence. Two people can communicate covertly by exchanging unclassified messages containing confidential information, but both parties have to take into account the presence of passive, active, or even malicious attackers.

The third person watching the communication should not be able to decide whether the sender is active in the sense that he sends covers containing secret messages rather than covers without additional information. More formally, if an observer has access to a carrier set $\{c_1,...,c_n\}$ of cover-objects transmitted between both communication parties, he should not be able to decide which cover-object c_i contains secret information. Thus, the security of invisible communication lies mainly in the inability to distinguish cover-objects from stego-objects [21,22].

In practice, not all data can be used as cover for secret communication, since the modifications employed in the embedding process should not be visible to anyone who is not involved in the communication process. This fact shows that sufficient redundant data should be contained in the cover for the purpose of being substituted by secret information. As an example, due to measuring errors, any data which are the result of some physical scanning process will contain a stochastic component called noise. Such random artifacts can be used for the submission of secret information. In fact, it turns out that noisy data has more advantageous properties in most steganographic applications.

Obviously a cover should never be used twice, since an attacker who has access to two versions of one cover can easily detect and reconstruct the message. To avoid accidental reuse, both sender and receiver should destroy all covers that already have been used for information transfer.

Some steganographic methods combine traditional cryptography with steganography: the sender encrypts the secret information prior to the embedding process. Clearly, such a combination increases the security of the overall communication process, as it is more difficult for an attacker to detect embedded Cipher text (which itself has a rather random appearance) in a cover. A strong steganographic system, however, does not need prior enciphering [21].

SIHT utilizes a secret key steganography system similar to a symmetric cipher: the sender chooses a cover c and embeds the secret message into c using a secret key k. If the receiver knows the key used in the embedding process, he or she can reverse the process and extract the secret message. Anyone who does not know the secret key should not be able to obtain the evidence of encoded information. Again, the cover c and the stega-objects can be perceptually similar.

A secret key steganography system can be described as a quintuple [153] $\Sigma = \langle C, M, K, D_K, E_K \rangle$, where C is the set of possible covers, M is the set of secret messages with $|C| \geq |M|$, K is the set of secret keys, $E_K : C \times M \times K \rightarrow C$ and $D_K : C \times K \rightarrow M$ with the property that $D_K(E_K(c,m,k),k) = m$ for all $m \in M$, $c \in C$, and $k \in K$.

9.3.5 SIHT OPERATING MODES

SIHT used for encryption speech information communication offers two working modes: transparent mode and secure mode. By just simply using a key to switch from one mode to the other, this process takes less than 10 seconds [86,87,90,148].

Transparent mode is a status where communication between speakers is established in the usual way. Secure mode is used when privacy is required. Any one of the speakers presses a key and the microcontroller leaves transparent mode, transmitting a call establishment sequence. When the other modem, set by default in originate mode, detects the sequence, it replies to complete the handshaking and reports the situation to its microcontroller to abandon, as well, the transparent mode. If modem handshaking has been successfully completed, information to generate the encryption key is transferred after the procedure. Once all the preliminary settings are completed, both acoustic and visual signals report to the users that secure mode is active. Subsequently, information is transferred securely and privately over the PSTN. If any problem arises and the data link is broken, acoustic and visual warnings will inform the users of the state of the system before returning to transparent mode.

The user interface informs the users of the state of the system in two ways: on a liquid quartz display and by voice messages. Messages are Stabilizing Communication, Interrupted Communication, Terminal Readies for Secret Communication, and others.

Each equipment holds a secret card reader. Only authorized cards may have access to the SIHT. Each card is customized according to the buyer organization requirements. This includes mainly the definition of a closed group of users.

9.3.6 ARCHITECTURE OF SIHT

In this section, the architecture of SIHT is presented [86,87,90].

9.3.6.1 Computational Load Distribution

Specific hardware architecture has been developed for SIHT. It is a single board containing four DSPs and one single-chip microcontroller. Four DSPs work in parallel: three of the DSPs (floating point TMS320C31) are used as slaves, implementing the speech coder, speech decoder, encryption algorithm, decryption algorithm, speech embedding algorithm, and speech extraction algorithm individually. Another fixed point DSP (TMS320C54x) is used as master, acting as communication controller, bus arbitrator, and interface circuit controller, interfacing with E.Modem. A single-chip microcontroller Intel 80C196 is used for interfacing with displayer, keyboard, and smart card reader. Master DSP TMS320C54x is incorporated with 80C196 executing bus control.

Task partition at the highest level divides the system into major blocks: E.Modem, speech coder, speech decoder, encryption, decryption, speech embedding, speech extraction, and man–machine interface.

The global architecture and flow chart of SIHT are described as follows.

9.3.6.2 Architecture Details

The architecture of SIHT is shown in Figure 9.11 [86,87,90].

The modem was implemented on an E.Modem module in a single configuration with the supporting standard AT command set and dial-up and auto-answer functions. After careful analysis of the vast information and taking the availability into account, the E.Modem type EM5600 was adopted for SIHT. This module is connected with master DSP TMS320C54x and single-chip microcontroller Intel 80C196. It is a V.90bis modem that operates full-duplex over two wires and interfaces in synchronous and asynchronous mode with a broad range of transmission rates. EM5600 has both serial and parallel interfaces with the controller, supporting a secondary channel at 150 bps, near and far echo cancel schemes, automatic adaptive equalizer, automatic rate negotiation, auto-test, and so on.

Speech coding is performed with a commercial vocoder based on CELP and ADPCM schemes. These approaches assure good reconstruction results but demand considerable computational resources, so floating point DSPs have to be used.

The speech coder, speech decoder, encryption algorithm, decryption algorithm, embedding algorithm, and extraction algorithm are implemented on three floating point DSPs (TMS320C31 @ 40 MFLOPS) and one fixed point DSP (TMS320C54x @ 40 MFLOPS). The three slave DSPs are external RAMless versions. The master DSP has 64 k words of 0 wait states SRAM for program and data [90]. Again, they are coupled tightly and the master has the slaves' memory in its own memory

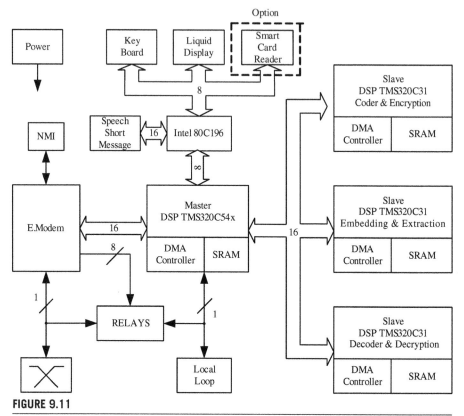

FIGURE 9.11

SIHT system architecture.

map. Arbitration is resolved by using hardware without protocol penalties. The only peripheral for this DSP set is another 16-bit codec for the local loop. The codec is connected to the telephone circuit via several relays governed by the modem. The greatest computational load is brought by the CELP algorithm. Encryption is supposed to be a simple subroutine of the speech process due to its compact structure. By examining all subtasks of the coder and the data flow requirements, the implicit sequential execution of the global process is observed. Only two of these subtasks are capable of parallelization: vector searching and pitch detecting. Both subtasks are separated from the large grain speech coder task and allocated in both the two slaves.

The equipment has a unique 4 Mbit ROM that contains the software for both modem and speech coder [148]. It is in charge of distributing software to the system devices at power up.

The man–machine interface consists of a two-line LCD display, a 16-key keyboard, and a groove for memory card insertion. All these elements are governed by an Intel 80C196 single-chip microcontroller and communications with the remaining blocks are via serial ports.

Special care has been taken of physical level communications between processors. Accesses to no-wait-state accompanied by very simple protocols discharges the processor from delays in message passing. An exhaustive organization of the software has been a fundamental topic in order to get a system without communications overhead. The DMA controller from DSP TMS320C54x devices permits a shared memory-like structure that saves hardware and eliminates arbitration complexity. Several routines of this kind of send–receive have been developed suitable for DMA facilities. FIFO structures, single word semaphores, and unidirectional interrupts (generated by communications slaves) sensibly speed up high-level protocols. Of course, the last ideas refer to access control rather than logical delays caused by semaphores or algorithm answer times.

9.4 SUMMARY

In this chapter, a PC-based secure communication system and SIHT have been described. They are secure communication systems based on the technique of information hiding to transmit secret speech over PSTN. In these systems, information hiding is a generic term meaning either public speech information hiding or secret speech information concealing.

The technology that is developed and designed for these systems is different from the traditional encryption technology. The secret speech information is hidden (embedded) in the public speech information, and then transferred. Encrypted speech information by the traditional method is a pile of meaningless noise that will attract the attention of an illegal eavesdropper. Eavesdroppers will analyze, compare, and decrypt the received signal. If they cannot decrypt the signal, they will interfere with and destroy the signal on their own initiative. This will cause the damnification of the transferring signal, the receiver cannot get the proper signal, and furthermore, the transmission will be interrupted, which will cause incalculable losses. The hidden secret speech information that is intercepted by the illegal eavesdropper via PSTN seems to be public speech information and cannot draw the attention of the illegal eavesdropper, so it is very easy to escape from the attack of the illegal eavesdroppers. This is the kernel of the information hiding technique, which is the basis for designing the PC-based secure communication system and SIHT.

SIHT has been tested with China PSTN; communication at 33.6 kbps has achieved more than 94% of the trials, with an excellent speech quality. SIHT can recognize complicated speakers' speeches and better understanding is guaranteed.

This encryption procedure is being tested to assure a high degree of privacy, and to avoid the danger of message understanding by accidental or undesired eavesdroppers, and to make it difficult for cryptoanalysts to decipher the conversation in case it is intercepted.

SIHT is a system based on information hiding techniques, providing secure communication over PSTN. It uses commercial DSP chips and is capable of ciphering speech.

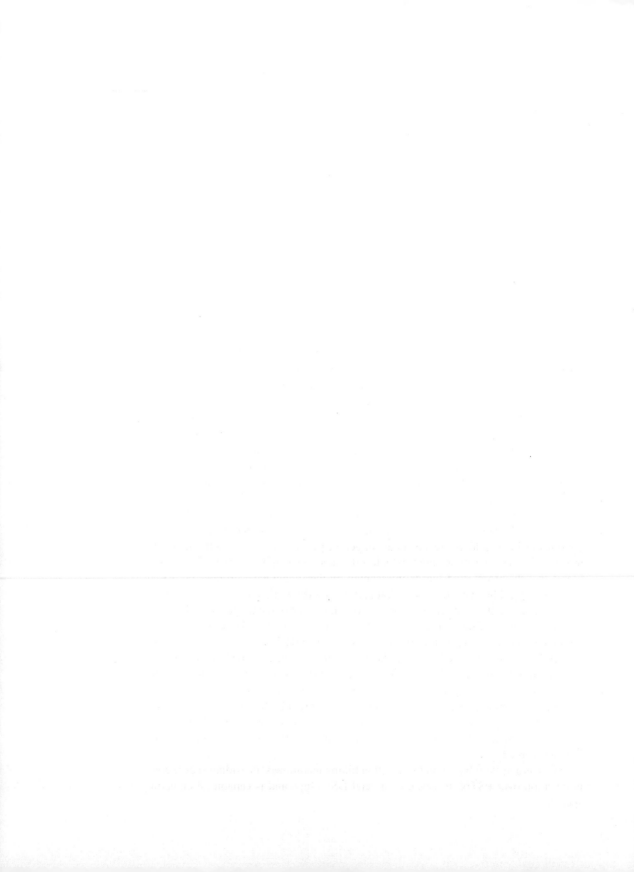

References

[1] Emmanuel Sodipo, The Art of Security and Information Hiding [M], Computers, Published: (Jun 15), 2011.

[2] M. Srivatsa, A. Iyengar, Liu Ling, Jiang Hongbo, Privacy in VoIP Networks: Flow Analysis Attacks and Defense, IEEE Transactions on Parallel and Distributed Systems 22 (4) (2011) 621–633.

[3] T. Ogunfunmi, M.J. Narasimha, Speech over VoIP Networks: Advanced Signal Processing and System Implementation, IEEE Circuits and Systems Magazine 12 (2) (2012) 35–55.

[4] A.D. Keromytis, A Comprehensive Survey of Voice over IP Security Research, IEEE Communications Surveys & Tutorials 14 (2) (2012) 514–537.

[5] Thomas Porter, Jan Kanclirz Jr., Brian Baskin, Practical VoIP Security. Syngress, June 10, 2006, ISBN-10:1597490601, ISBN-13: 978-1597490603.

[6] David Endler, Mark Collier, Hacking Exposed VoIP: Voice Over IP Security Secrets & Solutions by DavidEndler. McGraw-Hill Osborne Media, 2006. ISBN-10: 0072263644, ISBN-13: 978-0072263640.

[7] James F. Ransome, John Rittinghouse, Voice over Internet Protocol (VoIP) Security, 2005, ISBN: 978-1-55558-332-3.

[8] Theodore Wallingford, Switching to VoIP, Publisher: O'Reilly Media, June, 2005, ISBN-13: 978-0-596-00868-0, ISBN-10: 0-596-00868-6, Pages in Print Edition: 504.

[9] Matsunaga Akira, Koga Keiichiro, Ohkawa Michihisa, An Analog Speech Scrambling System Using the FFT Technique with High-Level Security, IEEE Journal on Selected Areas in Communications 7 (4) (May 1989).

[10] F. Huang, E.V. Stansfield, Time Sample Speech Scrambler Which Does Not Require Synchronization, IEEE Transactions on Communications 41 (11) (1993) 1715–1722.

[11] Ozge Koymen, John Morton, Scott Wilkinson, Digital Speech Encryption, Compression, and Transmission, May 13, 1996, pp. 18–551.

[12] Lin-Shan Lee, Ger-Chih Chou, A General Theory for Asynchronous Speech Encryption Techniques, IEEE Journal on Selected Areas in Communications 4 (2) (March 1986) 280–287.

[13] Fulong Ma, Jun Cheng, Yumin Wang, Wavelet Transform-based Analogue Speech Scrambling Scheme, Electronics Letters 32 (8) (1996) 719–721.

[14] S.M. Cherry, Is Hyperlinking to Decryption Software Illegal?, IEEE Spectrum 38 (8) (2001) 64–65.

[15] B. Goldburg, S. Sridharan, E. Dawson, Design and Cryptanalysis of Transform-based Analog Speech Scramblers, IEEE Journal on Selected Areas in Communications 11 (5) (1993) 735–744.

[16] Shujun Li, Chengqing Li, Kwok-Tung Lo, Guanrong Chen, Cryptanalyzing an Encryption Scheme Based on Blind Source Separation, IEEE Transactions on Circuits and Systems 55 (4) (2008) 1055–1063.

[17] Antonio Servetti, J.C. De Martin, Perception-based Partial Encryption of Compressed Speech, IEEE Transactions on Speech and Audio Processing 10 (8) (2002) 637–643.

[18] K. Li, Y.C. Soh, Z.G. Li, Chaotic Cryptosystem with High Sensitivity to Parameter Mismatch, IEEE Transactions on Circuits and Systems I: Fundamental Theory and Applications 50 (4) (2003) 579–583.

[19] Xia Yongxiang, C.K. Tse, F.C.M. Lau, Performance of Differential Chaos-shift-keying Digital Communication Systems over a Multipath Fading Channel with Delay Spread, IEEE Transactions on Circuits and Systems II: Express Briefs 51 (12) (2004) 680–684.

[20] Ning Jiang, Wei Pan, Lianshan Yan, Bin Luo, Shuiying Xiang, Lei Yang, Di Zheng, Nianqiang Li, Chaos Synchronization and Communication in Multiple Time-Delayed Coupling Semiconductor Lasers Driven by a Third Laser, IEEE Journal of Selected Topics in Quantum Electronics 17 (5) (2011) 1220–1227.

[21] F.A.P. Petitcolas, R.J. Anderson, M.G. Kuhn, Information Hiding – A Survey [J], IEEE Transactions of Proceedings of Theory 87 (7) (July 1999) 1062–1078.

[22] Moulin Pierre, A. Joseph, O'Sullivan, Information-theoretic Analysis of Information Hiding [J], IEEE Transactions on Information Theory 49 (3) (2003) 563–593.

[23] Jiagui Wu, ZhengMao Wu, Tang Xi, Fan Li, Wei Deng, GuangQiong Xia, Experimental Demonstration of LD-Based Bidirectional Fiber-Optic Chaos Communication, IEEE Photonics Technology Letters 25 (6) (2013) 587–590.

[24] Zhijun Wu, Wei Gao, Wei Yang, LPC Parameters Substitution for Speech Information Hiding, Journal of China Universities of Posts and Telecommunications 16 (6) (December 2009) 103–112.

[25] Zhijun Wu, Lan Ma, A Novel Approach of Secure Communication Based on the Technique of Speech Information Hiding, Journal of Electronics (China) 23 (2) (March 2006) 152–156.

[26] Rongyue Zhang, Vasiliy Sachnev, Magnus Bakke Botnan, Hyoung Joong Kim, Jun Heo, An Efficient Embedder for BCH Coding for Steganography, IEEE Transactions on Information Theory 58 (12) (December 2012).

[27] Hasan Mahmood, Tariq Shah, Hafiz Malik, Source Coding and Channel Coding for Mobile Multimedia Communication. Mobile Multimedia-User and Technology Perspectives (Chapter 5), Publisher: InTech, January, 2012. ISBN: 978-953-307-908-0, 156 pages.

[28] Randy Yates, A Coding Theory Tutorial, Digital Signal Labs, August 2009. http://www.digitalsignallabs.com888-708-3698.

[29] S. Roman, Coding and Information Theory. Springer, 1992.

[30] Tarik Zeyad, Ahlam Hanoon, Speech Signal Compression Using Wavelet and Linear Predictive Coding, Al-Khwarizmi Engineering Journal 1 (1) (2005) 52–60.

[31] Vijay Garg, Speech Coding and Channel Coding, Wireless Communications & Networking (Chapter 8), Publisher: Morgan Kaufmann, 13 June 2007, ISBN-13: 978-0-12-373580-5, ISBN-10: 0-12-373580-7.

[32] Zhigang Jiang, Sandeep K. Gupta, Threshold Testing: Improving Yield for Nanoscale VLSI, IEEE Transactions on Computer-aided Design of Integrated Circuits and Systems 28 (12) (December 2009).

[33] S. Tassart, Band-Limited Impulse Train Generation Using Sampled Infinite Impulse Responses of Analog Filters, IEEE Transactions on Audio, Speech, and Language Processing 21 (3) (2013) 488–497.

[34] S. Tassart, Time-Invariant Context for Sample Rate Conversion Systems, IEEE Transactions on Signal Processing 60 (3) (2012) 1098–1107.

[35] F. Pflug, T. Fingscheidt, Robust Ultra-Low Latency Soft-Decision Decoding of Linear PCM Audio, IEEE Transactions on Audio, Speech, and Language Processing 21 (11) (2013) 2324–2336.

[36] R. Nishimura, Audio Watermarking Using Spatial Masking and Ambisonics, IEEE Transactions on Audio, Speech, and Language Processing 20 (9) (2012) 2461–2469.

[37] B. Matschkal, J.B. Huber, Spherical Logarithmic Quantization, IEEE Transactions on Audio, Speech, and Language Processing 18 (1) (2010) 126–140.

[38] Xiguang Zheng, Ritz, C., Jiangtao Xi, Encoding Navigable Speech Sources: A Psychoacoustic-Based Analysis-by-Synthesis Approach. IEEE Transactions on Audio, Speech, and Language Processing, 21 (1) (2013) 29–38.

[39] C.O. Etemoglu, V. Cuperman, A. Gersho, Speech Coding with an Analysis-by-synthesis Sinusoidal Model. IEEE International Conference on Acoustics, Speech, and Signal (ICASSP'00), Istanbul, Turkey, 5–9 June 2000, Vol. 3, pp. 1371–1374.

[40] Chi-Sang Jung, Young-Sun Joo, Hong-Goo Kang, Waveform Interpolation-Based Speech Analysis/Synthesis for HMM-Based TTS Systems, IEEE Signal Processing Letters 19 (12) (2012) 809–812.

[41] Tokunbo Ogunfunmi, Madihally Narasimha, Principles of Speech Coding, CRC Press Inc, April, 2010, ISBN: 0849374286.

[42] Homer Dudley, R.R. Riesz, S.S.A. Watkins, A Synthetic Speaker, Journal of the Franklin Institute, June 1939, 227 (6) 739–764.

[43] Emre Gündüzhan, Kathryn Momtahan, A Linear Prediction Based Packet Loss Concealment Algorithm for PCM Coded Speech, IEEE Transactions on Speech and Audio Processing 9 (8) (November 2001).

[44] Y.J. Liu, James D. Mosko, Vector Predictive Coding for Very Low-Bit-Rate Speech Encoders, IEEE Military Communications Conference – Crisis Communications 2 (1987) 538–542.

[45] C. Crippa, G. Nicollini, et al. A 2.7-V CMOS Single-Chip Baseband Processor for CT2/CT2+ Cordless Telephones, IEEE Journal of Solid-State Circuits, IEEE Journals & Magazines 34 (2) (1999) 170–181.

[46] Angel M. Gómez, José L. Carmona, José A. Gónzález, Victoria Sánchez, One-Pulse FEC Coding for Robust CELP-Coded Speech Transmission Over Erasure Channels, IEEE Transactions on Multimedia 13 (5) (October 2011).

[47] John Willis O'Leary, B.Sc.E.E, A VLSI Architecture for Linear Prediction Analysis of Speech, Carletton University Ottawa, April 1989, ISBN: 0-315-51180-x.

[48] Series G: Transmission Systems and Media, Digital Systems and Networks, Digital Terminal Equipments – Coding of Voice and Audio Signals, Coding of Speech at 16 kbit/s Using Low-Delay Code Excited Linear Prediction, ITU-T Recommendation G.728. ITU-T, June, 2012.

[49] General Aspects of Digital Transmission Systems, Dual Rate Speech Coder for Multimedia Communications Transmitting at 5.3 and 6.3 kbit/s, ITU-T Recommendation G.723.1, ITU-T, Match, 1996.

[50] J.D. Gibson, M.G. Kokes, Data Embedding for Secure Communications. MILCOM 2002, Proceedings, California, US: 7–10 October 2002, Vol. 1, pp. 406–410.

[51] Series G: Transmission Systems and Media, Digital Systems and Networks, Coding of Speech at 8 Kbit/s Using Conjugate-structure Algebraic-code-excited Linear-prediction (CS-ACELP), ITU-T Recommendation G.729. ITU-T, Retrieved 2009-07-21.

[52] J.D. Gibson, Speech Coding Methods, Standards, and Applications, IEEE Circuits and Systems Magazine 5 (4) (2005) 30–49.

[53] Valentin Emiya, Emmanuel Vincent, Niklas Harlander, Volker Hohmann, Subjective and Objective Quality Assessment of Audio Source Separation, IEEE Transactions on Audio, Speech, and Language Processing 19 (7) (2011) 2046–2057.

[54] Luís Miguel, Malveiro Pereira, Tomaz Roque, Quality Evaluation of Coded Video[D], Universidate tecnica Lisboa, July, 2009.

[55] M. Winograd, Ilija Stojanovi, Eric Metois, Rade Petrovi, Kanaan Jemili, Joseph, Data Hiding Within Audio Signals. June 15, MIT Media Lab, Series: Electronics and Energetics, 12 (2) (2002) 103–112.

[56] Kiyoshi Tanaka, Yasuhiro Nakamura, Kineo Matsui, Embedding Secret Information into a Dithered Multi-Level Image, Department of Computer Science, The National Defense Academy, Yokosuka, 239 Japan Tel 0468(41)3810 ex. 2286.

[57] Abeer Tariq, Ekhlas Falih, Eman Shaker, A New Approach for Hiding Data within Executable Computer Program Files Using an Improvement Cover Region. IJCCCE, 13 (1) (July 2013).

[58] Tom Spring, IBM Updates Copy-Protection Software (currently available online at http://www.pcworld.idg.com.au/article/24205/ibm_updates_copy-protection_software/).

[59] Pamela Samuelson, The Uneasy Case for Software Copyrights Revisited, George Washington Law Review Arguendo 79 (6) (September 2011) 1746–1782.

[60] S. Katzenbeisser, F. A. P. Petitcolas. Information Hiding Techniques for Steganography and Digital Watermarking. Vienna and Cambridge, UK: Artech House, Inc., 2000, ISBN: 1-58053-035-4.

[61] Naofumi Aoki, Lossless Steganography for Speech Communications, Recent Advances in Steganography (Chapter 5), Edited by Hedieh Sajedi, Published by InTech, 2012, ISBN: 978-953-51-0840-5.

[62] N. Aoki, A Packet Loss Concealment Technique for VoIP Using Steganography Based on Pitch Waveform Replication, IEICE Transactions on Communications J86-B (12) (2003) 2551–2560.

[63] N. Aoki, A Band Extension Technique for G.711 Speech Using Steganography, IEICE Transactions on Communications E89-B (6) (2006) 1896–1898.

[64] N. Aoki, A Band Extension Technique for G.711 Speech Using Steganography Based on Full Wave Rectification, IEICE Transactions on Communications J90-B (7) (2007) 697–704.

[65] N. Aoki, A Technique of Lossless Steganography for G.711, IEICE Transactions on Communications E90-B (11) (2007) 3271–3273.

[66] N. Aoki, A Technique of Lossless Steganography for G.711 Telephony Speech, 2008 Fourth International Conference on Intelligent Information Hiding and Multimedia Signal Processing (IIHMSP2008), Harbin, China, pp. 608–611.

[67] N. Aoki, Lossless Steganography Techniques for IP Telephony Speech Taking Account of the Redundancy of Folded Binary Code, AICIT 2009 Fifth International Joint Conference on INC, IMS and IDC (NCM2009), Seoul, Korea, pp. 1689–1692.

[68] N. Aoki, A Lossless Steganography Technique for G.711 Telephony Speech, 2009 APSIPA Annual Summit and Conference (APSIPA ASC 2009), Sapporo, Japan, pp. 274–277.

[69] N. Aoki, A Lossless Steganography Technique for DVI-ADPCM, Transactions on Fundamentals of Electronics, Communications and Computer Sciences J93-A (2) (2010) 104–106.

[70] N. Aoki, A Semi-Lossless Steganography Technique for G.711 Telephony Speech, Sixth International Conference on Intelligent Information Hiding and Multimedia Signal Processing (IIHMSP2010), Darmstadt, Germany, pp. 534–537.

[71] N. Aoki, Enhancement of Speech Quality in Telephony Communications by Steganography, Multimedia Information Hiding Technologies and Methodologies for Controlling Data, IGI Global, 2012.

[72] Wojciech Mazurczyk, VoIP Steganography and Its Detection – A Survey, Computer Science Cryptography and Security, April 2013.

[73] W. Mazurczyk, Z. Koulski, New VoIP Traffic Security Scheme with Digital Watermarking, In Proceedings of the 25th International Conference on Computer Safety, Reliability and Security SafeComp 2006, Lecture Notes in Computer Science 4166, pp. 170–181.

[74] W. Mazurczyk, Z. Koulski, New Security and Control Protocol for VoIP Based on Steganography and Digital Watermarking, In Proceedings of the 5th International Conference on Computer Science – Research and Applications (IBIZA 2006b), Kazimierz Dolny, Poland.

[75] W. Mazurczyk, K. Szczypiorski, Covert Channels in SIP for VoIP Signaling. In Proceedings of the 4th International Conference on Global E-security, London, United Kingdom, 2008, pp. 65–70.

[76] W. Mazurczyk, K. Szczypiorski, Steganography of VoIP Streams. In : Meersman, R., Tari, Z., (eds) OTM 2008b, Part II – Lecture Notes in Computer Science (LNCS) 5332, Springer-Verlag, Berlin, Heidelberg, Proceedings on the Move Federated Conferences and Workshops: 3rd Internation Symposium Information Security (IS'08), Monterrey, Mexico, 2008, pp. 1001–1018.

[77] W. Mazurczyk, J. Lubacz, Lack – A VoIP Steganographic Method, Telecommunication System: Model Anal Des Manag 45 (2/3) (2010) 153–163.

[78] W. Mazurczyk, K. Cabaj, K. Szczypiorski, What are Suspicious VoIP Delays? Multimedia Tools and Applications 57 (1) (2010) 109–126.

[79] W. Mazurczyk, P. Szaga, K. Szczypiorski, Using Transcoding for Hidden Communication in IP Telephony, In: Computing Research Repository (CoRR), abs/1111.1250, arXiv. org E-print Archive, Cornell University, Ithaca, NY (USA) 2011, http://arxiv.org/abs/1111.1250.

[80] W. Mazurczyk, Lost Audio Packets Steganography: A First Practical Evaluation, International Journal of Security and Communication Networks, John Wiley & Sons, 2012, ISSN: 1939-0114.

[81] Y. Huang, B. Xiao, H. Xiao, Implementation of Covert Communication Based on Steganography, International Conference on Intelligent Information Hiding and Multimedia Signal Processing (IIH-MSP 2008), Harbin, China, pp. 1512–1515.

[82] Y. Huang, J. Yuan, M. Chen, B. Xiao, Key Distribution over the Covert Communication Based on VoIP, Chinese Journal of Electronics 20 (2) (2011) 357–360.

[83] Y. Huang, S. Tang, J. Yuan, Steganography in Inactive Frames of VoIP Streams Encoded by Source Codec, IEEE Transactions on Information Forensics and Security 6 (2) (2011) 296–306.

[84] Y. Huang, S. Tang, Y. Zhang, Detection of Covert Voice-over Internet Protocol Communications Using Sliding Window-Based Steganalysis, IET Communications 5 (7) (2011) 929–936.

[85] Y. Huang, S. Tang, C. Bao, Yau J. Yip, Steganalysis of Compressed Speech to Detect Covert Voice over Internet Protocol Channels, IET Information Security 5 (1) (2011) 26–32.

[86] Zhijun Wu, Research and Implementation for Speech Information Hiding, Journal of Communication 23 (8) (August 2002).

[87] Wu Zhijun, Design of Speech Information Hiding Telephone. IEEE Region 10 Annual International Conference, Proceedings of Tencon, 28–31 October 2002, Vol. 1, pp. 113–116.

[88] Zhijun Wu, Wei Yang, Yixian Yang, ABS-based Speech Information Hiding Approach, IEEE Electronics Letters 39 (22) (30 October 2003) 1617–1619.

[89] Zhijun Wu, Speech Information Hiding in G.729, Chinese Journal of Electronics (CJE) 15 (3) (July 2006) 545–549.

[90] Zhijun Wu, Yun Hu, Xinxin Niu, et al. Information Hiding Technique Based Speech Secure Communication over PSTN, Chinese Journal of Electronics 15 (1) (2006) 108–112.

[91] Zhijun Wu, W. Yang, G.711-Based Adaptive Speech Information Hiding Approach, In Proceedings of ICIC 2006, LNCS 4113, pp. 1139–1144.

[92] L. Ma, Zhijun Wu, W. Yang, Approach to Hide Secret Speech Information in G.721 Scheme, In Proceedings of ICIC 2007, LNCS 4681, pp. 1315–1324.

[93] M. Khodaei, K. Faez, New Adaptive Steganographic Method Using Least Significant-Bit Substitution and Pixel-Value Differencing, Image Processing, IET 6 (6) (2012) 677–686.

[94] C.H. Chuang, G.S. Lin, Adaptive Steganography-Based Optical Color Image Cryptosystems. IEEE Proceedings of International Symposium on Circuits and Systems (ISCAS'09), 24–27 May 2009, Taipei, China. New York, NY, USA, 2009, pp. 1669–1672.

[95] F.A.P. Petitcolas, R.J. Anderson, Weaknesses of Copyright Marking Systems. Proceedings of the 6th ACM International Multimedia and Security Workshop (Multimedia'98), 12−16 September 1998, Bristol, UK. New York, NY, USA: ACM, 1998, pp. 55–62.

[96] X.X. Dong, M.F. Bocko, Z. Ignjatovic, Data Hiding via Phase Manipulation of Audio Signals. IEEE Proceedings of International Conference on Acoustics, Speech and Signal Processing (ICASSP'04): Vol. 5, 17–21 March 2004, Montreal, Canada. Piscataway, NJ, USA: 2004, pp. 377–380.

[97] Rashid Ansari, Hafiz Malik, Ashfaq Khokhar, Data-Hiding in Audio Using Frequency-Selective Phase Alteration, Department of Electrical and Computer Engineering, University of Illinois at Chicago, Illinois, USA.

[98] Nguyen Dinh-Quy, Gan Woon-Seng, A.W.H. Khong, Time-Reversal Approach to the Stereophonic Acoustic Echo Cancellation Problem, IEEE Transactions on Audio, Speech, and Language Processing 19 (2) (2011) 385–395.

[99] D. Gruhl, W. Bender, A. Lu, Echo Hiding. Information Hiding: First International Workshop on Information Hiding, Lecture Notes in Computer Science, Springer 174 (1996) 295–315.

[100] O.T.C. Chen, W.C. Wu, Highly Robust, Secure, and Perceptual-Quality Echo Hiding Scheme, IEEE Transactions on Audio, Speech, and Language Processing 16 (3) (2008) 629–638.

[101] W.C. Wu, O.T.C. Chen, Y.H. Wang, An Echo Watermarking Method Using an Analysis-by-Synthesis Approach. IEEE Proceeding of the 5th IASTED International Conference on Signal and Image Processing (SIP'03), 13–15 August 2003, Honolulu, HI, USA. Piscataway, NJ, USA, 2003, pp. 365–369.

[102] Deng Qianlan, The Blind Detection of Information Hiding in Color Image, International Conference on Computer Engineering and Technology (ICCET) 7 (April 2010) 346–348.

[103] R. Ansari, H. Malik, A. Khokhar, Data-Hiding in Audio Using Frequency-Selective Phase Alteration. IEEE Proceedings of International Conference on Acoustics, Speech and Signal Proceeding (ICASSP'04): 17−21 March 2004, Vol. 5, Montreal, Canada. Piscataway, NJ, USA, 2004, 389–392.

[104] V.K. Munirajan, E. Cole, S. Ring, Transform Domain Steganography Detection Using Fuzzy Inference Systems. IEEE Proceedings of the 6th International Symposium on

Multimedia Software Engineering (MSE'04), 13–15 December 2004, Miami, FL, USA. Piscataway, NJ, USA, 2004, pp. 286–291.

[105] N.M. Charkari, M.A.Z. Chahooki, A Robust High Capacity Watermarking Based on DCT and Spread Spectrum. IEEE Proceedings of the 7th IEEE International Symposium on Signal Processing and Information Technology (ISSPIT'07), 15–18 December 2007, Cairo, Egypt. Piscataway, NJ, USA, 2007, pp. 194–197.

[106] C.E. Shannon, A Mathematical Theory of Communication, Bell System Technical Journal 27 (October 1948) 379–423.

[107] E. Shannon, Communication Theory of Secrecy Systems, Bell System Technical Journal 28 (4) (1949) 656–715.

[108] J. Zöllner, H. Federrath, H. Klimant, A. Pfitzmann, R. Piotraschke, A. Westfeld, G. Wicke, Wolf, Modeling the Security of Steganographic Systems. 2nd Workshop on Information Hiding: April 1998, Portland, LNCS 1525, Springer-Verlag, 1998, pp. 345–355.

[109] Christian Cachin, An Information-Theoretic Model for Steganography. 2nd Workshop on Information Hiding: April 1998, Portland, LNCS 1525, Springer-Verlag, 1998, pp. 306–321.

[110] N.J. Hopper, J. Langford, L. van Ahn, Provably Secure Steganography, In Advances in Cryptology, Lecture Notes in Computer Science, Springer 2442 (2002).

[111] Suvarup Saha, Information Theoretic Characterization of Information Hiding, lecture notes, course project: Topics in Information Theory, EECS Department, Northwestern University.

[112] T. Mittelholzer, An Information-Theoretic Approach to Steganography and Watermarking in Information Hiding, 3rd International Workshop, IH'99 (A. Ptzmann, ed.), Lecture Notes in Computer Science, Springer, 1999, Vol. 1768, pp. 1–16.

[113] Daimao Lin, Lan Hu, Yunbiao Guo, Zhou S Linna, The Mechanism and Model of Generalized Information Hiding Technology, Journal of Beijing University of Posts and Telecommunications 28 (1) (2005) 1–5.

[114] Lan Ma, Zhijun Wu, Yun Hu, Wei Yang, An Information-Hiding Model for Secure Communication. ICIC (1) 2007, LNCS 4681, pp. 1305–1314.

[115] G.J. Simmons, The History of Subliminal Channels, IEEE Journal on Selected Areas in Communication 16 (4) (1998) 452–462.

[116] Shun Dong Li, Zheng Qin, A Kind of Subliminal Channel Communication Algorithm with Keys, Chinese Journal of Computers 26 (1) (January 2003).

[117] J.N. Daigle, Queueing Theory for Telecommunications, Addison-Wesley Publishing Company (February 1999), pp. 37–41.

[118] Chen Liang, Xiongwei Zhang, Speech Hiding Algorithm Based on Speech Parameter Model, Chinese Journal of Computers 26 (8) (August 2003) 954–981.

[119] Yibao Yu, Shuozhong Wang, Cluster Property Analysis and Experimental Evaluation of Mutual Information Measure for Speech Recognition, Signal Processing 18 (5) (October 2002) 442–447.

[120] Xiaohui Meng, Ling Xiao, Jie Cui, Multichannel Speech Enhancement System Based on Hearing Perceptual Model, Computer Engineering 36 (13) (2010) 9–12.

[121] W. Bender, D. Gruhl, N. Morimoto, et al. Techniques for Data Hiding [J], IBM System Journal 35 (3–4) (1996) 313–336.

[122] Han Jiqin, Zhang Lei, Zheng Tieran, Speech Signal Processing, Tsinghua University Press, 2004.

[123] F. Thomas, Quatieri. Discrete-Time Speech Signal Processing: Principles and Practice, Prentice Hall, 2001.

[124] N.M. Anas, Z. Rahman, A. Shafii, M.N.A. Rahman, Z.A.M. Amin, Secure Speech Communication over Public Switched Telephone Network. Applied Electromagnetics, 2005. Asia-Pacific Conference on Selangor, December 2005, ISBN: 0-7803-9431-3.

[125] Sviatoslav Voloshynovskiy, Frederic Deguillaume, Oleksiy Kovall, Thierry Pun. Information-Theoretic Data-Hiding for Public Network Security, Services Control and Secure Communications. Proceeding of Telsiks 2003, Serbia and Montenegro, NIS, 1–3 October 2003, pp. 3–17.

[126] S.K. Pal, P.K. Saxena, S.K. Mutto, The Future of Audio Steganography. Pacific Rim Workshop on Digital Steganography, Japan, 2002.

[127] P. Moulin, J.A. O'Sullivan, Information-Theoretic Analysis of Watermarking, I Proceedings of 2000 IEEE International Conference on Acoustics, Speech, and Signal, Istanbul, Turkey, 6 (June 2000) 3630–3633.

[128] Series G: 32 kbit/s adaptive differential pulse code modulation (ADPCM). ITU-T Recommendation G.721. ITU-T, 1988.

[129] Series G: General Aspects of Digital Transmission Systems, Pulse Code Modulation (PCM) of Voice Frequencies, ITU-T Recommendation G.711. ITU-T, 1988.

[130] M.U. Celik, G. Sharma, A.M. Tekalp, E. Saber, Lossless Generalized-LSB Data Embedding, IEEE Transactions on Image Processing 14 (2) (February 2005) 253–266.

[131] A.M. Kondoz, Digital Speech: Coding for Low Bit Rate Communication Systems, John Wiley & Sons, West Sussex, England, 2004.

[132] C. Wai, Chu, Speech Coding Algorithms-Foundation and Evolution of Standardized Coders, John Wiley & Sons, Inc, Hoboken, New Jersey, 2003.

[133] Zhijun Wu, Haixin Duan, Xing Li, An Approach to Hide Secret Speech Information, Journal of Shanghai Jiaotong University, Science E-11 (2) (2006) 134–139.

[134] Liang Chen, Xiongwei Zhang, Study of Information Hiding in Security Speech Communication, Journal of PLA University of Science and Technology 3 (6) (2002) 1–5.

[135] Wang Bingxi, Speech Coding. Xi'dian Press, 2002, pp. 1–256.

[136] S.M. El Noubi, M. El-Said Nasr, E.S. Gemeay, Performance of Analysis-by-Synthesis Low-Bit Rate Speech Coders in Mobile Radio Channel. In Proceedings of the Nineteenth National Radio Science Conference (NRSC 2002), Alexandra, Egypt, 19–21 March 2002, Vol. 1, pp. 363–371.

[137] ETSI, Digital Cellular Telecommunications System (Phase 2+), Full Rate Speech, Transcoding (GSM 06.10 version 7.0.0 Release 1998), pp. 10–59.

[138] Hu Licai, Wang Shuozhong, Information Hiding Based on GSM Full Rate Speech Coding, IEEE, 2006.

[139] Rui Miao, Yongfeng Huang, An Approach of Covert Communication Based on the Adaptive Steganography Scheme on Voice over IP. Proceedings in IEEE International Conference on Communications, 2011, pp. 1–5.

[140] Tian Hui, Study on Cover Communication Based on Streaming Media [D], Huazhong University of Science and Technology, June 2010.

[141] Hatada Mitsuhiro, Sakai Toshiyuki, Komatsu Naohisa, Yamazaki Yasushi, A study on digital watermarking based on process of speech production. Information Processing Society of Japan, Special Interest Group notes, 2002, No. 43, pp. 37–42.

[142] Yamin Su, Yongfeng Huang, Xing Li, Steganography-Oriented Noisy Resistance Model of G.729a. IMACS Multiconference on Computational Engineering in Systems Applications, October 2006, Beijing, China.

[143] Cheng Yimin, Guo Zhichuan, Xie Chunhui, Xie Yuming, Covert Communication Method Based on GSM for Low-Bit-Rate Speech, Journal of Circuits and Systems 13 (2) (April 2008).

[144] S. Zander, G. Armitage, P. Branch, Covert Channels and Countermeasures in Computer Network Protocol, IEEE Communication Magazine 45 (12) (2007) 136–142.

[145] S.H. Sellke, C. Wang, S. Bagchi, N. Shroff, TCP/IP Timing Channels: Theory to Implementation. In: Proceedings of the 28th IEEE International Conference on Computer Communications, Rio de Janeiro, Brazil, 2009, pp. 2204–2212.

[146] A. Westfeld, F5-A Steganographic Algorithm. Proceedings of the 4th International Workshop on Information Hiding, Lecture Notes in Computer Science 3137 (2001) 289–302.

[147] Hae Y. Yang, Kyunng H. Lee, Sang H. Lee, Method and Apparatus for Partially Encrypting Speech Packets. United States Patent Publication, Feburary 2009.

[148] L. Díez-del-Río, S. Moreno-Pérez, et al. Secure Speech and Data Communication Over the Public Switching Telephone Network, IEEE Transactions on Proceedings of Theory 48 (8) (April 1994) 425–428.

[149] N.F. Johnson, S. Jajodia, Exploring Steganography: Seeing the Unseen, IEEE Transactions on Computer of Theory 31 (2) (February 1998) 26–34.

[150] Zhijun Wu, Lan Ma, Xiaoyun Shen, Video conference technology development and examples in Visual C++ [B], Posts & Telecommunications Press, Beijing, China, January 2006.

[151] Lan Ma, Xiaoyun Shen, Di Wan, Master the technologies of video / audio encoding and decoding in Visual C++ [B], Posts & Telecommunications Press, Beijing, China, July, 2008.

[152] American National Standards Institute, TIA/EIA-232-F, ANSI, 1997.

[153] D. Denning, Cryptography and Data Security, Addison-Wesley Publishing Co, Reading, Mass, 1982.

[154] B.S. Atal, V. Cuperman, A. Gersho, Advances in Speech Coding, Kluwer Academic, 1991.

Index

175

Printed and bound by CPI Group (UK) Ltd, Croydon, CR0 4YY

03/10/2024

01040327-0009